THE END OF PHYSICS

THE END OF PHYSICS

The Myth of a Unified Theory

DAVID LINDLEY

BasicBooks
A Division of HarperCollinsPublishers

Designed by Ellen Levine

93 94 95 96 ◆ / H C 9 8 7 6 5 4 3 2 1

Library of Congress Cataloging-in-Publication Data
Lindley, David.
   The end of physics : the myth of a unified theory / David Lindley.
      p.   cm.
   Includes bibliographical references and index.
   ISBN 0–465–01548–4
   1. Grand unified theories (Nuclear physics) 2. Physics—Philosophy.
I. Title.
QC794.6.G7L55 1993
539.7'2—dc20   92-54524
                                                            CIP

CONTENTS

ACKNOWLEDGMENTS

GABRIELE PANTUCCI, my agent, and Susan Rabiner, at Basic Books, provided inestimable amounts of early support and enthusiasm, which on more than one occasion compensated for temporary weariness on my part. Sara Lippincott did a sterling job of editing the final manuscript.

I am grateful to Lisa Heinz for many discussions, for the use of her library, and for comments on an early draft. My colleagues at *Nature*, especially Laura Garwin, were admirably forgiving of the time and attention this book took away from my real job.

Prologue

The Lure of Numbers

IN 1960, the Hungarian-American physicist Eugene Wigner published an essay entitled "The Unreasonable Effectiveness of Mathematics in the Natural Sciences."[1] This was by no means the plaintive lament of an aging scientist who couldn't cope with the increasing mathematical sophistication and complication of modern physics: Wigner himself was a distinguished theoretical physicist, who in the 1940s had encouraged his colleagues to take up new kinds of mathematics, the theory of symmetries in particular. What nagged at Wigner was the evident fact that mathematics is invented by mathematicians for quite their own reasons and purposes, yet seems inexorably to find its way into theories of the physical sciences. This transfer had become increasingly noticeable in the twentieth century, as physicists found places in their theories for the mathematics of curved geometries, symmetry relations, matrix algebra, and the like—concepts produced during the heyday of Victorian science and regarded at that time as the purest of pure mathematics. What puzzled Wigner puzzled many others, including Albert Einstein: "How can it be that mathematics," he once asked, "being a product of human thought which is independent of experience, is so admirably appropriate to the objects of reality?"[2]

Modern mathematical science did not spring into existence overnight, but a convenient marker for its birth is the publication, in 1687, of Isaac Newton's *Philosophiae naturalis principia mathe-*

matica, the Mathematical Principles of Natural Philosophy. Newton's *Principia* was a great catalog of laws, enunciated in mathematical form, including his three famous laws of motion and all manner of deductions from them. Before Newton, science was a collection of vague ideas and wishful thinking. Bodies fell to the ground, according to Aristotle, because it was their natural tendency to do so, a formula that strikes us today as little more than a tautology. Newton mathematized physics to turn the "tendencies" and "potentials" of Aristotelian thought into what we now recognize as true science, based on quantitative and precise relations between masses, forces, and motions. Numbers were needed to measure these quantities, so physical laws had to take the form of relationships between numbers—mathematical laws, in other words. As the scope of science grew, and as scientists sought not merely to depict nature but to understand it, the mathematical form of natural laws began to take on an importance of its own. In the nineteenth century, the Scottish physicist James Clerk Maxwell was able to combine the laws governing electricity and magnetism into a unified theory of electromagnetism, which was prized for its mathematical tidiness as well as for its ability to account for physical phenomena. At the beginning of the twentieth century, Einstein produced his special and general theories of relativity, which likewise made elegant use of mathematics to describe the geometry of space and the rules of gravitational attraction. The shape of these theories, the mathematical and geometrical purity of them, were inherently persuasive. Even today direct experimental tests of general relativity have been few, and what still draws the student of physics to the theory is its seemingly inborn appeal, the sense that something so profoundly satisfying must necessarily contain a deep truth.

Maxwell put electricity and magnetism together into a single theory of electromagnetism; Einstein combined gravity with geometry to create general relativity; more recently, in the 1970s, Sheldon Glashow, Steven Weinberg, and Abdus Salam were able to tie electromagnetism to something called the weak interaction, which governs certain kinds of radioactivity, thereby creating a unified electroweak interaction. There are now only three fundamental forces of nature—gravity, the electroweak interaction, and the strong nuclear force, which holds protons and neutrons together in the atomic nucleus. Theoretical physicists today have some ideas on how to combine the strong and electroweak forces, and hope eventually to include gravity in a single unified theory of all the forces. The very meaning of "unification" in this context is that one harmonizing theoretical structure, mathematical in form, should be made to accommodate formerly distinct theories under one roof. Unification is the theme, the backbone of modern physics, and for most physicists what unification means in practice is the uncovering of tidier, more comprehensive mathematical structures linking separate phenomena. The question posed by Wigner and Einstein—Why should physics be inherently mathematical?—is still there, at the bottom of all that has been done in physics in this century.

After the 1920s, physicists were trying to understand and extend the newly created theory of quantum mechanics, and were beginning to open up the world of nuclear physics. These were occupations that would keep them busy for many decades. But Einstein was moving in a different direction. He did not think that electromagnetism and gravity should remain separate, and spent the later years of his life searching for a theory of "electro-gravity" to complete the work begun by Maxwell. It is generally reckoned now that nothing much came of his efforts, but his ambition was surely correct. With electroweak unification more or less a finished job, theoretical physicists have moved on to grand unification, which includes the strong force. Grand unification remains a somewhat tentative accomplishment, theoretically attractive but with no real evidence in its favor. Even so, physicists are trying to push beyond it. The remaining task is to include gravity in the grand-unification scheme, and this, more or less, is what Einstein was attempting earlier in the century. We can see now that Einstein was too far ahead of his time: gravity, in the modern view, can be combined with electromagnetism only after electromagnetism has been joined to the weak and strong forces.

Because his interests diverged so far from the mainstream of physics, Einstein established no real intellectual lineage. He is revered, but unlike Niels Bohr and Wolfgang Pauli and Werner Heisenberg and the other founders of twentieth-century physics, he was never a father figure to researchers of a new generation. His numerous attempts to find a theory of electrogravity led nowhere, and have not been pursued by others. Perhaps his inability to proceed toward the theory he desired led him to question the fundamental article of scientific faith—that there is indeed a theory to be found. He began to marvel that it is possible to do science at all, expressing his puzzlement at one time by saying, "The world of our sense experiences is comprehensible. The fact that it is comprehensible is a miracle."[3]

These are not the worries of a young, confident, eager scientist. Scientists in their prime forge ahead without anguishing over the philosophical soundness of their efforts. They simply assume that there must be a rational answer to the questions they are asking and a logically coherent theory that will tie together the phenomena they are studying. If scientists did not believe this—if they thought that at bottom the physical world might turn out to be irrational, to have some irreducible fortuitous element in it—there would be no point in doing science. When scientists begin to wonder, as Einstein and Wigner did, how science is possible at all, which is ultimately what their questioning of the mathematical basis of science is about, they are searching for reassurance, for some proof that there really is a fundamental theory out there in the dark waiting to be hunted down.

There is a temptingly simple explanation for the fact that science is mathematical in nature: it is because we give the name of science to those areas of intellectual inquiry that yield to mathematical analysis. Art appreciation, for example, requires a genuine effort of intellect, but we do not suppose that the degree to which a painting touches our emotions can be organized into a set of measurements and rules. There is no mathematical formula that, applied to the colors and shapes of a picture, produces some measure of its effect on us. Science, on the other hand, deals with precisely those subjects amenable to quantitative analysis, and that is why mathematics is the appropriate language for them. The puzzle becomes a tautology: mathematics is the language of science because we reserve the name "science" for anything that mathematics can handle. If it's not mathematical to some degree at least, it isn't really science.

Many sciences make use of mathematics, but only theoretical physics is thoroughly steeped in it. Late in the nineteenth century, it seemed that the comprehensive system we now call classical physics, dealing with motions and forces and electromagnetism and heat and light and optics, might be nearing completion. Almost everything was understood, and the mathematical basis for that understanding was the differential equation, which above all encapsulates the ideas of flows and motions and rates of change. But the completion of classical physics was forestalled by the emergence of quantum theory and relativity, which introduced the notions of discontinuous change and curved space. In the twentieth century, such otherworldly, seemingly abstract mathematical notions as probability waves, symmetries, and matrix algebra have been incorporated into physics. The preeminence of the differential equation in classical physics can be rationalized by saying that it is the single piece of mathematics designed to cope with the principles on which classical physics is based, but what puzzled Einstein and Wigner, in the post-classical era, is the surprising usefulness of all manner of arcane pieces of mathematical knowledge.

Physicists, however, are scavengers of mathematics; they take what they need, adapt it to their purposes, and discard the rest. There are areas of mathematics that have occupied the attentions of mathematicians for thousands of years—for instance, all the theory that has grown up since the days of Euclid to account for the distribution of prime numbers—which have not yet found any application in physics. And although modern theoretical physics has taken over much of the language and ideas of group theory and symmetry to arrange the elementary particles and their interactions, it is also true that the full sophistication of the subject, as developed by mathematicians, is more than the physicists have taken advantage of.

It becomes hard to know if the question asked by Einstein and Wigner is profound or trivial. It is not true that all of science is mathematical, and it is not true that all of mathematics is used in

science. The odd thing, perhaps, is that as physics has reached farther, both into the microworld of elementary particles and into
the macroworld governed by general relativity, it has made use of
increasingly recondite mathematical ideas. Why should it be that
in "fundamental" physics, that arm of science which presumes to
grapple with the true and ultimate elements of nature, fancy
kinds of mathematics keep turning up? This is not a scientific
question; it is like asking why nothing can travel faster than the
speed of light. Theoretical physicists can invent theories in which
the speed of light is not an absolute limit, but those theories do
not correspond to the world we inhabit. The speed of light does
not have to be finite, but in our world, as distinct from all the
imaginary worlds a mathematician might invent, it is. Some
things in the end can be determined only empirically, by looking
at the world and figuring out how it works. Asking why one kind
of theory should turn out to describe nature while other, equally
plausible, equally attractive theories do not is moving beyond the
kind of question that physics is fit to answer.

Max Planck, the reluctant originator of quantum mechanics,
observed that "over the entrance to the gates of the temple of science are written the words: Ye must have faith."[4] Underpinning
all scientific research, but especially research in fundamental
physics, is this article of faith: that nature is rational. Elementary
particles are finite objects, and they interact in specific ways; their
masses, charges, and other quantum properties, and the forces
between them, are fixed and calculable. What physicists essentially want to do is to attach numbers to physical quantities and
find relationships between those numbers. This is how physics is
supposed to work, and so firm is this notion that one cannot
really imagine any alternative. In a science such as botany, for
example, a good degree of qualitative description can be accepted,
because it is understood that individual plants are enormously
complex biological constructions whose characteristics are not
supposed to be predictable from first principles, in terms of a
handful of fundamental botanical precepts. But the whole point
of fundamental physics is precisely that the elementary particles
are presumed to be, at some level, finite and indivisible objects,
completely described by a handful of numbers and equations.

To abandon the notion that physics ought to reduce ultimately to a set of elementary particles and a description of the way they interact would be to abandon an intellectual tradition going back to the very beginnings of science; it would be to abandon science. But if a description of elementary particles and interactions is indeed the goal, then mathematics ought to provide the language needed. Mathematics is, after all, the totality of all possible logical relations between quantitative objects (numbers, sets of numbers, functions relating one set of numbers to another), and if we believe that fundamental physics is likewise a set of relations—not all possible relations, just those that actually obtain in the real world—among elementary particles with fixed properties and characteristics, then it seems inevitable that the appropriate language to describe physics is indeed mathematics. Therefore, Einstein's and Wigner's question—How can it be that mathematics is the correct description for the objects of reality?—is really the same as Einstein's much more profound statement that the comprehensibility of the universe is a miracle.

Putting all these metaphysical worries aside, and granting that mathematics is indeed the language of science, we can see that there are two extremes in the way scientific research can be conducted. One way is to assemble as many facts and as much data as seem relevant to the problem at hand, in the hope that the appropriate mathematical relations will pop out, or at least will be perceived by the trained and receptive scientific mind. The other way is to find mathematical laws whose beauty and simplicity have some particular appeal, and then attempt to fit the world to them. The latter method, which seems rather rarefied to us nowadays, is what Pythagoras intended, millennia ago, by his idea of "the harmony of numbers." He and his followers believed that the world ought to be based on mathematical concepts—prime numbers, circles and spheres, right angles and parallel lines—and their method of doing science was to strain to understand the phenomena they observed around them in terms of these presupposed concepts. It was for many centuries a central belief that the whole world, from the motion of falling bodies to the existence of God, could ultimately be understood by pure thought alone. Thus philosophers and theologians labored mightily to construct "proofs" of the existence of God; thus thinkers such as Aristotle and Plato declared that nature worked according to certain principles, and sought to deduce from those principles the phenomena of the natural world, relegating to a distant second place empirical tests of whether the world behaved as their theories required it to.

The revolutionary idea behind modern science—the science of Copernicus, Kepler, Galileo, and Newton—was that facts took on new importance. The fundamental belief that theories should be simple and harmonious was retained, but the notion that these theories could be developed by cogitation alone was abandoned. Philosophers, too, starting with John Locke, began to recognize the independent importance of empirical facts—facts not deducible by any method of pure reason. The huge success of science in the last three centuries is owed to the adoption of an essentially pragmatic philosophy. Science cannot be deduced from first principles devoid of empirical content, but neither can it be inferred from an uncritical inspection of the facts. The creative element in the scientist's mind is essential: inspecting the facts, the mind of the scientist, replete with mathematical relationships and constructs, perceives dimly the form of a theory that might work. With this first guess, the process of deduction, experimental test, refinement, and further hypothesis can take off, and science is built up. In all this there must be a sort of equality, a balance, between facts and theories. Beautiful theories are preferable to ugly ones, but beauty alone does not make a theory correct; facts are to be collected and taken note of, but the judicious scientist knows when to ignore certain contradictory facts, realizing that a tentative theory cannot explain everything and anticipating that all will be well in the end.

The contention of Nicolaus Copernicus that the Sun is at the center of the Solar System and that the planets orbit around it illustrates how science walks forward on two legs. His scheme was meant to supplant the Ptolemaic system, wherein the Earth was at the center and the Sun and planets traveled around it in epicycles (circles upon circles), which had been fantastically elaborated over the centuries in order to explain the observed motions of the celestial bodies. Copernicus' first intent was simplification: if the

Sun was at the center and the planets moved around it, the whole system became simpler and therefore, perhaps, more likely to be true. But he held on to one piece of received knowledge, which was that the orbits had to be circular. This was an aesthetic choice, going back to the Greeks, who thought that circles and spheres were touched by the divine while other shapes were base.

But the Copernican system, with the planets following perfectly circular orbits around the Sun, actually worked less well, in terms of reproducing accurately the observed movement of the planets, than the vastly more complex but highly evolved Ptolemaic system. Copernicus was attacked by the Church chiefly for dogmatic reasons, because he removed the Earth from its central position in the universe, but it should be understood that his model was in fact not all that good. What his idea had going for it, however, was intellectual appeal and beauty: his model was simple and elegant. This simplicity of conception was not easy to dismiss, and a few years later, by the expedient of using elliptical orbits rather than circles (another defeat for dogma), Johannes Kepler showed how the Copernican model of the Solar System could accurately reproduce the observed motions of the planets. This was its final triumph.

There was an interim when the two systems competed, the inelegant accuracy of the Ptolemaic against the beauty but relative inaccuracy of the Copernican, and this contest is a model of the progress of science. The scientist always strives for accuracy, but there are times when the desire for accuracy is a hobble that must be cast off to allow pursuit of a new and appealing idea. This dilemma pervades science, and according to the state of science and the mood of the times scientists will line up on one side of the debate or the other. Provided experiments can be done, science will move forward in its time-honored zigzag way, with theorists, entranced by some new vision of mathematical beauty, moving off in some direction only to be pulled back when a practical test of the theory fails, forcing a reexamination of its principles and motives.

Scientific history, however, tends not to record the zigzags. History is written by the winners, and failed attempts to explain this or that phenomenon are soon forgotten, though the failed theories may have provided a useful stimulus. Moreover, scientists have a habit of covering their tracks when they publish. Kepler spent years working with the observational data collected by the Danish astronomer Tycho Brahe, and trying who knows how many geometrical shapes for the planetary orbits before he hit on ellipses. But when he published his results, ellipses were all that he offered the reader. In the same way, Isaac Newton, though he presumably reached his theories of motion and gravity and his theorems on orbital shapes by some sort of trial and error (albeit guided by general principles), wrote up his researches in the *Philosophiae naturalis principia mathematica* as if he had been fortunate enough to glimpse directly the divine rules that govern the universe and, thus blessed, had worked out the consequences. Newton remains a paragon of the scientific style for theoretical physicists. He starts with grand pronouncements, an ex-cathedra air about them, and proceeds to demonstrate their success in explaining various observed phenomena. The *Principia* was set out with vatic authority, and illustrated with theorems and examples to convince the reader of the fundamental truths therein displayed. The principles were elegant, the examples majestically undeniable, and all was well.

Ultimately, theories must be useful and accurate, and in that sense it does not matter whether they were arrived at through piercing insight or blind luck. A theory may come into the world mysteriously and inexplicably, but once formulated it is subject to minute rational analysis and test. Experiment weeds out theories that work from those that don't. But the more sophisticated the theories become, and the more delicate the facts they try to explain, the longer this process of weeding out will take. And during the weeding out, aesthetic considerations are not misplaced. The Copernican system was attractive before it was accurate; had it not been attractive it would never have succeeded. Aesthetic considerations are not a blemish on the scientific method but an essential part of it. Sometimes it is not just useful but absolutely necessary to put aesthetic considerations ahead of the mere facts, and scientists are not foolish to pursue appealing ideas that do not with perfect accuracy account for the facts.

This all assumes, of course, that theories can eventually be tested. What restrains the theorist from becoming wholly carried away by the attractions of some mathematical theory is the need to make predictions with it and to test them against the hard realities of the real world. But as, during this century, experiments in fundamental physics have become harder to do, more costly, and more consuming of time and manpower, this restraining empirical influence has been weakened. At the same time, it is apparent that the greatest successes of twentieth-century physics have been theories that make use of novel and enticing mathematical structures imported from the realm of pure thought. Aesthetic judgments are taking on greater importance in theoretical physics not as the result of some conscious change in the methods of science but by default. The constraint of experiment on the scientific imagination is growing weaker, and in some areas is almost absent. In such circumstances scientists may have nothing but aesthetics—a nose for the truth—to guide them.

Some of the blame, unfortunately, for this shift back toward the old Pythagorean ideal must go to Albert Einstein. His general theory of relativity is the prime example of an idea that convinces by its mathematical structure and power, and for which experimental verification is something of an afterthought. This is an exaggeration, of course: an early prediction of Einstein's theory was that gravity should bend the paths of light rays away from straight lines, and in 1919 the British astronomer Sir Arthur Eddington led a team to the tropics to measure the bending of light from several stars as they passed behind the Sun during a total eclipse. His successful measurement was hailed, both in the popular press and within the scientific community, as proof that Einstein's theory was correct. Nevertheless, general relativity remains even today one of the least well tested of physical theories. It has passed all tests to which it has been put, and it is more elegant than any rival theory of gravity, but its predominant position in physics rests largely on its power to connect in a coherent and beautiful theoretical framework a handful of experimentally reliable facts.

Einstein became more and more fond, as he grew older, of using aesthetic justification as a criterion for scientific correctness. When asked what he would have thought if Eddington's expedition had not found the bending of light by the Sun, he said, "Then I would have been sorry for the dear Lord; the theory is correct." The dear Lord was Einstein's stand-in for aesthetic principles in science; he meant by the phrase something that more old-fashioned commentators might have called "Mother Nature." "God is subtle, but He is not malicious" was his general opinion of the way science should be; scientific theories should be elegant, and could well be complex and hard to fathom, but they should not be contrived; they should possess some inner beauty marking their aptness to describe nature. Of the new rules in quantum mechanics, which appeared to mean that some essential part of nature was determined not absolutely but only in a statistical or probabilistic way, Einstein remarked that in his view God did not play dice. That was his rationale for objecting to the foundations of quantum mechanics, and for insisting that some more funda-mental theory must lie below its surface; he could not accept that the dear Lord would allow a probabilistic theory to be the whole truth about the way the world works. And in connection with his attempts to understand how the universe came about and what determined its appearance today, Einstein said once that "what interests me is whether God had any choice in the creation of the universe." He was commenting on the perfectly reasonable scien-tific question of how our particular universe, from all possible universes allowed by physics, came into being, but the hope expressed here was that somehow, when all the correct theories of physics were known and understood, only one universe would be possible. As always, he wanted a strict form of determinism to be at the bottom of the most fundamental questions, and, as always, he wanted the arbiter of this scientific determinism, the judge of which universe is the right and only one, to be "God"—by which he meant not some divine being who picks a universe, as if from a shop window, according to whim, but a guiding prin-ciple, an aesthetic power that can allow only what is right, ele-gant, and proper to come into being and survive.

Einstein was, most of the time, a man of profound humility, and perhaps it was all right for him constantly to defer to his sense of aesthetic judgment, or "God." He was not nearly as cov-etous of his ideas as most thinkers are, and he was prepared to give up an idea if it transgressed his interpretation of the way the world ought to work. But in his later years, when he was tinkering with one mathematical construction after another in his search for a unified theory of electromagnetism and gravity, he seems to have lost his inner compass and drifted aimlessly on the seas of mathematical invention. The danger of the Einsteinian style to science is that most physicists have never possessed his inner compass in the first place but use his methods as an excuse for playing at mathematics, constructing theories that look good but are essentially empty of deeper justification.

The lure of mathematics is hard to resist. When, by dint of great effort and ingenuity, a previously vague, ill-formed idea is encapsulated in a neat mathematical formulation, it is impossible to suppress the feeling that some profound truth has been discovered. Perhaps it has, but if science is to work properly the idea must be tested, and thrown away if it fails.

Not all of science is mathematical. Not everything mathematical is science. Yet there persists in theoretical physics a powerful belief that a sense of mathematical rigor and elegance is an essential, perhaps overriding, element in the search for the truth. For theoretical physicists, the lure of mathematics is more sophisticated than a simple predilection for dealing with numerical quantities and ignoring what cannot be quantified; there is the pure aesthetic appeal of mathematical ideas and systems in themselves, regardless of any applicability to the real world. Playing with mathematics and numerical relationships has led to many a true scientific discovery. Quantum mechanics was born when Max Planck, struggling to understand the relationship between the temperature of a glowing body and the color and intensity of the light it emitted, played at fitting mathematical formulas to experimentally derived results until he found one that worked. Only afterward did he realize that the successful formula contained a novel implication: the energy in light must come in little packets—quanta—not in a continuous flow. In the same vein, Niels Bohr began to understand how the quantum idea could explain the characteristic frequencies of light emitted by hydrogen atoms because he was familiar with a simple numerical formula, stumbled across many years before by the Swiss mathematician Johann Jakob Balmer, which expressed all the known wavelengths of hydrogen light in terms of the difference between the reciprocals of the squares of whole numbers. Balmer's formula had been the result of arithmetical guesswork, but it turned out to be of profound significance.

There are many cases, however, where an excessive devotion to the search for mathematical simplicity can mislead. For example, there is a venerable numerical formula that claims to explain the distances of the planets from the Sun by simple arithmetic. This idea, now known as Bode's law, after the eighteenth-century German astronomer who promulgated it, holds that the distances form a series based on successive powers of two (2, 4, 8, 16, and so on). In fact, the distances do not exactly follow this law, and there are discrepancies: where Bode's law puts the fourth planet there is only a collection of chunks of rocks called the asteroids, and Pluto, the outermost planet, does not fit the formula at all. Then again, the asteroids may be the fragments of a former planet, and Pluto, an odd globe in many respects, may be an interloper, captured after the Solar System was formed. Although Bode's law is clearly neither correct nor complete, there is enough in it to make it a source of wonder. Ought we in fact to expect the planetary distances to display some sort of arithmetic regularity? In a vague way, perhaps we should: if we believe that the planets formed slowly from a disk of dust and rocks circling the new-formed Sun, we might suspect that the physical laws of gravitational aggregation and cooling and adhesion of dust grains might lead to some sort of systematic ordering of the planets.

On the other hand, perhaps not. Bode's law is one possible mathematical relation out of many, and it is not accurately obeyed. How is one to judge, from the bare observational facts, whether the law is significant or not? In some other planetary system, in which the planetary distances were arranged quite differently, would not some other equally ingenious astronomer have come up with some other equally appealing mathematical law, approximately true, that hinted at the workings of some deeper principle in the disposition of the planets than mere happenstance?

This is the bane of empirical science. Ingenious minds, at work on a small set of numbers, can very often come up with beguilingly simple mathematical laws and relationships that really mean nothing at all. There is a thriving occupation on the fringes of science that goes by the name of numerology: Bode's law is the archetypal numerological relationship. Nowadays amateurs of science play around with the masses of the hundreds of subnuclear particles found at accelerators (the Particle Data Group, based at the Lawrence Berkeley Laboratory, in California, provides a convenient catalog of particle properties, updated every year or so) and come up with numerological relationships that "explain" the masses with formulas that tend to involve prime numbers, or the value of pi, or the square root of two, or some such. Almost all this activity is perfectly silly; it is the result of playing with numbers, and that is all it is.

Yet numerology contains an element of usefulness that makes it hard to dismiss the whole subject. Balmer's formula for the wavelengths of light from hydrogen atoms was, as noted, a pure exercise in numerology, but it became the basis for Bohr's quantum theory of the atom. More recently, theoretical physicists have settled on the quark model as an explanation for the constitution and internal structure of many subnuclear particles, but the quark model arose when physicists noticed certain patterns and groupings in the masses and other properties of newly discovered particles—patterns whose origin and meaning were obscure but which were too evident to be ignored. Arranging particle masses according to simple formulas is exactly what the modern physicist derides when it is done by amateurs, but when it was done by professionals it was the first step on the road to quarks.

Good and bad numerology are easily distinguished in retrospect. Bad numerology is the search for simple numerical relationships between quantities for which no simple physical relationship is thought to exist. Bode's law is bad numerology, because we cannot reasonably expect a messy process like the formation of planets to produce, after all the dust has settled, a piece of high-school arithmetic. Good numerology, on the other hand, is the search for simple numerical relationships that derive from simple underlying physical connections. Balmer's numerology did not by itself indicate what sort of new physics lay hidden in the hydrogen atom, but it was the key that opened up atomic structure to Bohr's insight.

What forms does numerology take today, and how are we to judge whether they are good or bad? In quantum electrodynamics, the quantum-mechanical version of Maxwell's electromagnetic theory, a number called the fine-structure constant repeatedly shows up. It is a combination of some other elementary physical numbers, including the speed of light and the charge on the electron, and it has the value $\frac{1}{137}$—almost. This beguiling value has attracted professional as well as amateur numerologists. Sir Arthur Eddington, the tester of general relativity, drifted off into areas of metaphysics and mysticism as he grew older, and in a book modestly called *Fundamental Theory* produced a famously incomprehensible explanation for the value of the fine-structure constant. But Richard Feynman, one of the inventors of quantum electrodynamics, also believed that the value of the fine-structure constant ought to be explicable in some basic way, although he was wise enough to venture no guesses.[5] If reputable physicists find a number like $\frac{1}{137}$ fascinating enough to look for greater meaning in it, we may think that there is useful numerology here, but until such time as a physical theory comes along to explain the number, our puzzling over the matter is not much use. In the end, numerological relationships must be counted among the many clues that scientists work from in their attempts to move theory along. History is the sole judge of whether they choose wisely which clues to attend to and which to ignore.

Lately there have been intimations that fundamental physics is nearing that desirable day when all is understood, when all the elementary particles of the world and all the forces and interactions that affect them will have been compiled into a single catalog, logically set out, with a place for everything and everything in its place. This can realistically be called the end of physics. It will not mean that computer chips have been made as fast as they can be made, that no stronger metal alloy will ever be devised, that lasers are as powerful as they are ever going to be. Even less will it mean that biochemistry and the nature of life will be understood.

There will be a lot more physics to do, not to mention more chemistry and biology, after the end of physics, but it will no longer be fundamental. The rules of the game will be known, like the rules of chess, and what remains will be to work out all the variations—a task that may well demand as much intellectual effort as figuring out the rules in the first place.

Now is not the first time that physicists have thought the end might be imminent. A hundred years ago, classical physics seemed to be approaching a state of perfection that offered to contain an explanation for every phenomenon in the natural world; but there were a couple of flaws in classical physics, and out of those flaws sprang quantum mechanics and the theory of relativity. Classical physics was the physics of atoms perceived as miniature billiard balls, interacting with each other by means of tangible forces, such as electrical and magnetic attraction. But quantum mechanics turned the imaginary billiard balls into tiny, fuzzy patches of uncertainty, and general relativity turned paper-flat space into curved space, and time into something variable and personal instead of universal and constant. Through the intervening hundred years, physicists have occupied themselves with working out quantum mechanics and relativity in all their implications, and in the process physics has absorbed mathematical ideas of increasing sophistication. Old-fashioned arithmetical numerology, even Feynman's daydreaming over the fine-structure constant, is not reckoned a worthy task. But physicists still search for mathematical harmonies—not simple arithmetical rules but structures that house symmetries and unities of logical form. The unification of the fundamental forces, which has been the goal of theoretical particle physics for decades, is exactly the search for a mathematical system containing within it the apparently different strong, weak, electromagnetic, and gravitational forces. Electroweak unification, the first step in this endeavor, has made some predictions that have been experimentally verified, but grand unification, which includes the strong force, has been favored by no empirical evidence and may never be so favored. Why then the continuing drive for unification, to include gravity as well?

Unified theories look better, so the thinking goes, than a set of separate theories. The mathematical structures describing unification have a certain elegance and power: it is remarkable that physical interactions so evidently different as the weak, electromagnetic, and strong forces can be made to appear as different aspects of the same thing. The achievement of grand unification is a mathematical tour de force. But is it any more than that? Does it lead to prediction and test, in the traditional way? Are physicists truly laying bare the fundamental laws of nature in these overarching mathematical schemes, or is the beauty of unification entirely in the eyes of the beholders?

The modern version of the end of physics looks quite different from the classical version. Nineteenth-century physicists believed that physics would be complete when every natural phenomenon could be described in essentially mechanical terms, by means of pictures based on forces and masses and rods and pulleys and strings. These pictures were to be taken quite literally; to the classical physicist, the fact that a physical phenomenon could be represented in such homely terms was the very indication that truth had been uncovered. But the modern physicist has done away with naïve mechanical models and believes that truth is to be found in logical, mathematical structures, no matter how bizarre and uncommonsensical they seem, no matter whether we will ever be able to perceive the purity of the structure directly. The most recent speculation of the theoretical physicists is that elementary particles are not particles at all but vibrations of tiny loops of quantum-mechanical string, wriggling around in twenty-six-dimensional space. This is the modern equivalent of the classical physicist's hope that all matter could be understood in terms of atoms that behaved essentially like little billiard balls. If strings rippling through twenty-six-dimensional space turn out to be the basis for our ultimate understanding of nature, the end of physics will hardly resemble the neat mechanical vision of classical days.

Modern particle physics is, in a literal sense, incomprehensible. It is grounded not in the tangible and testable notions of objects and points and pushes and pulls but in a sophisticated and indirect mathematical language of fields and interactions and wavefunctions. The old concepts are in there somewhere, but in heavy disguise. To the outsider, it may seem that the theoretical physicists of today are in the grip of a collective mathematical zaniness, inventing twenty-six-dimensional spaces and filling them with strings out of obfuscatory glee. Their use of language is as esoteric and baffling as that of the literary deconstructionists: they seem to speak in words and sentences, but it is a kind of code. The mathematical physicist and the deconstructionist share the same popular image: each speaks in a private gobbledygook understandable only to those similarly initiated. It is easy to imagine that both are lost in a world of pointless fabulation.

In one sense, such criticism of modern theoretical physics is philistine. Physics may be complex, mathematical, and arcane, but it is not capricious. The inventors of strings and twenty-six-dimensional spaces did not think up these things at random, simply to give themselves a new set of toys. There is a line of rational thinking that leads from the billiard-ball atoms of classical physics to the intangible mathematical entities of today. Physics is complicated because the world is complicated. It is the honest intent of physicists to find the simplest explanation they can for all the phenomena of the natural world, and it is a measure both of their ingenuity and of the complexity of the physical world that they have gone to such extremes of theorizing but still have not found a single theory that explains it all.

On the other hand, the inexorable progress of physics from the world we can see and touch into a world made accessible only by huge and expensive experimental equipment, and on into a world illuminated by the intellect alone, is a genuine cause for alarm. Even within the community of particle physicists there are those who think that the trend toward increasing abstraction is turning theoretical physics into recreational mathematics, endlessly amusing to those who can master the techniques and join the game, but ultimately meaningless because the objects of the mathematical manipulations are forever beyond the access of experiment and measurement.[6] When physicists come up with a new mathematical construction in which different pieces of fundamental physics are incorporated, it is natural enough for them to think they have achieved a further step toward complete unification and passed another milestone on the road to the end of physics. But what is the use of a theory that looks attractive but contains no additional power of prediction, and makes no statements that can be tested? Does physics then become a branch of aesthetics?

If there is an end of physics, in short, can we ever get there, and how will we know we have arrived? If the ultimate theory deals in entities that are, by their very construction, irreconcilably divorced from the measurable world, does it make any sense even to think that there is an ultimate theory? Will the end of physics be reached not because we have discovered all the answers but because the questions that remain are forever unanswerable, so that we must simply make do with what we have?

I

THE
MICROWORLD

THE WORLD is made of atoms; everyone knows that. But a hundred years ago the existence of atoms was so controversial that it was a matter for philosophical as well as scientific debate. Critics were alarmed that at the foundation of physics the atomists wanted to install objects that were far too small to be seen; it seemed to them that atoms were mental creations, not physical facts, and that if physics were built out of such things the whole of science was in danger of becoming an unverifiable fantasy. But that barrier has long been overcome. The debate was won by the atomists because, first, the atomic hypothesis proved extraordinarily fruitful, and second, experimental methods to detect and manipulate individual atoms were devised. We now know in a direct, verifiable way that atoms contain tiny nuclei, around which electrons orbit, and we know further that the nuclei are themselves little clusters of protons and neutrons. The atomic theory today is more than simply useful—it is experimentally demonstrated.

For the great majority of working scientists—chemists, biologists, physicists who are trying to make stronger metal alloys or faster silicon computer chips or more powerful lasers—the atom is the bottom rung of the ladder. But there is a small minority of physicists engaged in what is called particle physics, or elementary-particle physics, or fundamental physics. For these people neutrons and protons are giant, complex structures. The playground of the particle physicists is at a truly elemental level; they live among quarks and gluons and photons and muons and all the other oddly named inhabitants of the subnuclear universe. There is a remarkable and wide division between these special physicists and the rest of the scientific world. The majority of scientists treat atoms as ground zero, and all other knowledge, all other physical structures, are seen as springing from them. But the particle physicist starts at ground zero and works down. Neutrons, protons, and electrons are not the objects from which explanation is built but the objects to be explained. The goal of particle physics is to understand atoms in terms of more truly fundamental precepts, and thus to understand all of physics—and, implicitly, all the rest of science—completely.

For the scientists working at the surface, the practical significance of the exploration below ground is essentially nil. Nothing that the fundamental physicist will ever find out is going to change the known properties of neutrons, protons, and electrons; neither their existence nor their basic attributes are matters for debate. The quest in fundamental physics is above all an intellectual one. If the particle physicists one day find a complete, profound, and satisfying explanation for the existence and properties of neutrons, protons, and electrons, the work of chemists and biologists will not suddenly become easy. Particle physicists have occasionally got themselves into trouble for calling their work fundamental, as if their efforts alone embodied the true spirit of science and all else were trivial. But even the detractors cannot deny that the urge to dig deeper for a final truth is irresistible.

In their efforts to explore the subnuclear world, where they hope one day to hit rock bottom and discover a truth beneath which nothing more is to be found, particle physicists have inevitably drifted away from the tangible world of our everyday senses. They have not only split the atom but also built bigger and bigger machines to split it into smaller and smaller pieces. The structure they have uncovered in this microworld has turned out to be far more complicated than anyone present at the beginning of twentieth-century physics would have guessed. The microworld is not a simple place, and physicists have therefore not been able to keep their theories of it simple.

The progress from atomic physics to nuclear physics to elementary-particle physics has taken science into a world accessible only by means of large, complicated, and expensive experimental devices. Accelerators like the Tevatron at the Fermi National Accelerator Laboratory, near Chicago, can fling particles together at a million times the energy of those emerging from natural radioactive decay, and collisions between particles at these energies create, if only for a fleeting moment, all manner of normally hidden inhabitants of the subnuclear microworld. The Superconducting Supercollider, a machine twenty times more powerful than the Tevatron, is now being built a few miles south of Dallas. Near Geneva, the Europeans may soon build an accelerator of comparable size. These machines will let the particle physicists peer a little farther into the microworld—but not much farther. Accelerators more powerful than these are not practicable, and how physicists will continue to make progress in particle physics remains to be seen.

So far in this century, theory and experiment in particle physics have moved forward more or less in tandem. Theoretical understanding of the microworld has come fitfully, and the theories that now stand as the best and most complete effort to understand it are certainly incomplete, and intellectually appealing only in parts. If the theories are ungainly and full of mysterious adjustments, it is because experimental facts force them to be that way; a theory must first and foremost reflect the way the world is. Particle physicists, however, now stand at a threshold: they want to continue their theorizing, but they are running out of ways their theories can be tested. Physics has moved from the simple, now almost quaint notions of classical physics to the confusing and counterintuitive ideas of quantum mechanics and relativity, and on to the search for a "theory of everything." The present state of particle physics may seem ramshackle, but it is necessarily so. As their empirical knowledge of the microworld has increased, physicists have had to enlarge and adjust their theories accordingly. Physics has reached its present state for good reasons, and through the systematic application of scientific methods. How long it can keep this course we do not yet know.

1

Lord Kelvin's Declaration

DURING the second half of the nineteenth century, the study of physics in Great Britain was dominated by William Thomson, later Lord Kelvin. As a young man he was famous for his mathematical abilities, and during his career he turned his attention, seemingly at will, to all aspects of the physical world: astronomy, mechanics, electricity and magnetism, heat, geophysics. With barely a glance at the efforts of others, he would assess vastly complicated problems and set them straight. His style was to tie together disparate facts and bits of attempted theorizing into neat, logical packages; his work extended not only the compass of science but its orderliness. On top of his endeavors in physics, Thomson became a Victorian entrepreneur, first working out the mathematical theory governing the transmission of electric signals through cables across the Atlantic, then calculating the most efficient and cost-effective cable design, and finally organizing a commercial consortium to obtain ships and lay cable from Ireland to Newfoundland. For this successful enterprise he was made Sir William Thomson, and his elevation to the peerage in 1892 owed as much to his friendships with political and business leaders of the day as to his scientific accomplishments.

Kelvin's urge to set the entire house of science in order led him, in the mid 1850s, to turn his attention to the geologists. Their recent ideas had been bothering him. They were beginning to understand the slow processes of rock formation and erosion, and to appreciate how geological change, acting over long periods, formed the continents and the mountains. But in all their new hypothesizing they took it for granted that the arena of the past, in which geological change was played out, was for practical purposes infinite. It was Kelvin's assertion, on the other hand, that the world could be no more than one hundred million years old, a figure he obtained by applying the newly formulated principles of energy conservation and thermodynamics to the present state of the Earth and the Sun. In this matter he opposed not only the geologists but also, later, the biologists. Charles Darwin and his followers wanted more time than Kelvin would allow for natural selection and evolution to raise the human race from its lowly beginnings, and in the first edition of the *Origin of Species* Darwin attempted to match Kelvin, calculation against calculation, by estimating the age of the Weald, a geological feature in southern England, from contemporary estimates of the rate of erosion. He produced a figure of three hundred million years, which he found satisfactory to his purposes, but his method was roundly attacked by Kelvin and other physicists, and in the second edition of the book Darwin acceded to their superiority and sheepishly withdrew this number in favor of something closer to Kelvin's.

The controversy over the age of the Earth was noisy enough to be heard beyond the walls of science. The American historian Henry Adams—a friend of Clarence King, the founder of the US Geological Survey and a participant in the debate—referred to the "cossack ukase"[1] issued by Kelvin, and Mark Twain concluded that "as Lord Kelvin is the highest authority in science now living, I think we must yield to him and accept his views."[2] For both Adams and Twain, amateur observers of science, the weight of certainty seemed to rest with Kelvin and the laws of physics. Kelvin's arguments indeed seemed irrefutable, even to those who lacked the knowledge to follow them in detail. As to the age of the Earth, he observed merely that the planet was hotter inside (down deep mineshafts, for example) than at the surface, and deduced directly that heat must be flowing outward from the interior and radiating away into space. Since the Earth was losing heat, it must have been hotter in the past, and by assembling some simple physical arguments with available data on the properties of rocks and the interior temperature of the Earth at different depths, Kelvin proved to the satisfaction of physicists that the Earth must have been molten no more than a hundred million years ago. Neither life nor continents could exist on a molten planet, so that was all the time the geologists and biologists could be allowed.

The Sun's age he estimated with similar directness. No one knew at that time the source of the Sun's heat, but the most efficient and abundant source of energy that Kelvin could imagine was gravity itself. Originally he had proposed a theory in which meteorites were continually crashing onto the Sun's surface, where their energy of motion (which came from the Sun's gravitational pull) turned into heat. This theory did not get by the astronomers, who noted that the supply of meteorites Kelvin needed should be observed between the Earth and the Sun but was not. Kelvin modified his theory along lines suggested by the German physicist Hermann von Helmholtz, another pioneer of nineteenth-century physics. There was no need to have extra mass raining down onto the Sun, Kelvin realized, because the material of the Sun itself was being compressed by its own gravity; the Sun must be gradually shrinking, and this steady squeezing converted the Sun's gravitational energy into heat. But the process clearly could not go on forever. If the Sun's own mass was its only source of power, Kelvin calculated that it, like the Earth, could be no more than a hundred million years old. He recognized that there might be other sources of heat, such as chemical reactions in the solar interior, but compared with gravitational contraction these other processes would yield much less heat per unit of solar material, so their presence or absence would have no real effect on the age Kelvin fixed upon.

The geologists of the nineteenth century chafed at these restrictions, but in the end had to yield to them. A few were sufficiently skilled in physics to quibble with some of Kelvin's assumptions and estimates and win a few million years more for their side, but there was nothing they could do to markedly alter the numbers. This in fact had a salutary effect on geology. Since the publication of Charles Lyell's *Principles of Geology* in the early 1830s, geologists had adhered almost unthinkingly to an assumption that became known as "uniformitarianism," according to which all geological

change proceeded at a constant rate. Since erosion and rock formation were seen in the middle of the nineteenth century to be
slow processes, it followed that geological change had always
been slow. Armed with this idea, geologists thought of the past as
essentially infinite, and used it extravagantly. It was this extrava-
gance that spurred Kelvin to the attack; against the casual
assumptions and loose calculations of geology, the physicists of
the mid-nineteenth century could set some stern rules and tested
principles.

But the achievements of the physicists were themselves only
recently won. Today heat and energy are viewed almost self-
evidently as the the same kind of thing, but at the beginning of
the nineteenth century they were regarded quite differently.
Energy meant specifically the idea of energy invented a century
and half earlier by Isaac Newton; it was the quantity possessed by
solid bodies in motion. A flying cannonball had a lot of energy, a
falling snowflake had little. Heat, on the other hand, was mysterious. It was thought of as a sort of substance, perhaps an invisible
fluid, that could be transferred from one body to another but
could also be created apparently out of nothing, as when a drill bit
bores into metal. Was the heat fluid released from the pores of the
metal's structure by abrasion?

Then again, heat could be converted into energy, as in a steam
engine. As the Industrial Revolution gathered pace, steam engines
and other kinds of machinery were invented and improved by
engineers and craftsmen working largely in ignorance of scientific
ideas, for there was at first no science of steam engines. But science caught up. In 1824, the French physicist Sadi Carnot published a short essay, "Réflexions sur la puissance motrice du feu,"
which for the first time made a quantitative link between the heat
supplied to a steam engine and the amount of work that could be
got out of it. Carnot introduced the idea of efficiency, and showed
that the conversion of heat to energy depended on the extremes
of temperatures between which the engine cycled as it heated
water into steam and condensed it back into water again. Thus
was born the science of thermodynamics—the dynamics of heat.
The first law of thermodynamics states that in any physical
process the total amount of energy is conserved, though it may be

converted from one form to another: this seems today more or less a matter of definition, but in the early days the many different forms that energy could take had not yet been identified, and figuring out how much mechanical work was equivalent to some quantity of heat was not a simple matter. The second law of thermodynamics says that heat will flow only from hot to cold, which seems as obvious today as saying that water will run only downhill. But in these two laws the scientists of the early nineteenth century found subtle implications. Engineers were constantly tinkering with steam engines to get a little more effort out of the same amount of fuel, but, as Carnot first appreciated, there was a limit to what could be done. A certain amount of heat would always be wasted—given up to the surroundings and therefore lost as a source of mechanical work.

One of the most profound ideas of thermodynamics was advanced by the German physicist Rudolf Clausius, who in pondering the ratio of the heat wasted to the heat converted into mechanical work came up with a definition of the "useful," or available, energy in any system. He invented a new word—entropy—which is now common parlance, although its meaning is often garbled. A system with low entropy has a lot of energy that can be turned into work, and Clausius showed that the second law of thermodynamics was equivalent to the stricture that entropy must always increase. Heat will always pass from a body at high temperature to one at a lower temperature, and the flow of heat can be made to work an engine of some sort. But as heat flows, temperature is equalized and no more work can be obtained. The entropy of the final system, two bodies at the same temperature, is higher than it was for the system of one hot and one cold body. As the Industrial Revolution ran along, heat was turned into mechanical work at an accelerating rate, and the entropy of the world increased correspondingly.

The science of thermodynamics grew more sophisticated, and all kinds of physical systems were brought under its rule. Chemical reactions could both generate heat and use it up, so there had to be an interconversion between heat and some sort of internal chemical energy possessed by the reacting substances. Simple changes of form—ice melting into water, water boiling into

steam—involved flows of heat, so thermodynamics was connected with the physical structure of solids and liquids. Electrical
currents could heat wires, and chemical reactions in batteries
could generate electricity, so there must be an electrical form of
energy that could both be turned into other forms and be created
from them. Long before there was any detailed understanding of
chemistry, electricity, and the atomic structure of materials, thermodynamics set limits on what could be done with all the new
kinds of energy.

The drawing together of these various physical systems under
one roof took decades, and profoundly changed science. Formerly
independent subjects—heat, mechanics, chemistry, electricity—
were becoming parts of a much bigger whole, rooms in the same
great mansion. The construction of that mansion was the apex of
nineteenth-century science, and represented the culmination of a
certain scientific style, quantitative and precise. In the two centuries since Galileo and Newton inaugurated modern science by
replacing quaint notions of the "natural tendencies" of objects to
fall or float or attract or repel with exact definitions of energy and
force, their successors had followed the same model in all areas of
science. Laws of electrical and magnetic attraction had been precisely formulated; chemistry no longer consisted solely of catalogs
of reactions and affinities but was beginning to be governed by
strict laws of combination and association.

The coming together of the piecemeal achievements of previous centuries into the great physical system of the nineteenth
century was the foundation for Kelvin's attack on the geologists.
Making the Earth's history infinite was as inconceivable as making a perpetual-motion machine. If geologists wanted the Earth to
remain pleasantly warm (not cold, not hot) indefinitely, they
were casting aside the laws of thermodynamics. Heat had to come
from somewhere; it could not simply be assumed into existence.
Especially, Kelvin believed, not by geologists. By the late nineteenth century, these attacks had produced some effects. Many
geologists were gentleman scholars of the old fashion, unable to
cope with Kelvin's physical reasoning or his mathematics, but
some understood that his objections could not be wished away.
They convinced the rest of the profession that if geologists wanted

to look like serious scientists—more like physicists, that is—then they had to be serious about chronology and prove that their science could accomplish its aims in the time permitted.

This meant, ultimately, the abandonment of uniformitarianism, the philosophy that all geological change happens, and has always happened, at the same slow pace. Geologists were forced instead to suppose that episodes of rapid change had built the Earth, that the past had on occasion been quite different from the present. Mountain ranges must have been born in spasmodic convulsions of the Earth's surface, settling afterward into the quiet middle age we now perceive. It was a great struggle for geologists to fit the history of the Earth into the mere hundred million years granted them by Kelvin; but they managed to do the job, more or less, and in so doing they were forced to make their hypotheses about the development of the Earth more precise, more rational. For this they had some reason to thank Lord Kelvin.

The power of the new science of thermodynamics was something remarkable: without knowing anything about the internal mechanics of a steam engine, a physicist could put a number to the maximum efficiency it could achieve, just as Kelvin, without needing to speculate on the nature of the Earth's heat, could calculate the maximum age of the planet. The evident universality of thermodynamic reasoning seemed to lend physics an invincible ascendancy over other sciences. Without knowing the intricacies of steam engines or the nature of geological erosion or the causes of biological change, the physicist could nevertheless make authoritative pronouncements on all those subjects.

This aura of invincibility grew with the creation of a new branch of mathematical physics known as statistical mechanics, which we owe largely to the efforts of Ludwig Boltzmann, a versatile Austrian physicist, and J. Willard Gibbs of Yale University, one of the first American physicists to gain wide repute. In separate but parallel efforts, Boltzmann and Gibbs embarked on an ambitious program to derive the inexorable laws of thermodynamics from fundamental principles. Their starting point was the atomic hypothesis, that all matter is made of tiny indivisible units of some sort. This was not a new idea—various Greek philosophers had embraced atomism thousands of years ago—but neither was it widely accepted science. Atoms were individually unde-tected, perhaps undetectable, and the debate over whether or not matter was really atomic in nature seemed to many scientists more philosophical than scientific.

Boltzmann and Gibbs, at any rate, were atomists, and their idea was to use the atomic hypothesis to provide a theoretical under-pinning for thermodynamics. The heat energy of a volume of steam, for example, was to be understood as the total energy of motion of its constituent atoms; the pressure exerted by the steam was the result of atoms banging against the walls containing it. If more heat energy was put into the steam, its atoms would move around faster, bang into the walls with greater impact, and there-fore exert more pressure. The simple mechanical view of heat as the motion of atoms instantly did away with all the strenuous but fruitless attempts to portray heat as a kind of immaterial fluid or substance. Heat cannot be isolated from matter, because matter consists of atoms and heat is nothing more than the motion of those atoms. Heat is equivalent to energy because it *is* energy, the energy of bodies in motion.

But statistical mechanics went further than this. The frustrating laws of thermodynamics, which dictated that heat and energy would be dissipated in certain ways regardless of the engineer's skill, could be seen now as an inevitable consequence of the atomic nature of heat. In a steam engine, for example, the general idea is to transfer the heat energy of steam in a cylinder to a pis-ton, which then turns a crankshaft. It is easy to see why the speeding atoms push on the piston, but it is now also easy to see why the transfer of energy from steam to piston is inefficient. For the entire heat energy of the steam to be passed to the piston, all the atoms would have to bang into the piston at exactly the same time and stop dead in their tracks, leaving the steam devoid of atomic motion, and therefore absolutely cold. This, one can see, is not likely to happen.

Then again, it is not quite an impossibility. If the motion of atoms inside the cylinder of a steam engine is truly random, then it is conceivable that in one particular steam engine somewhere in the Victorian world, all the atoms would happen, by chance, to have just the right disposition so that they would indeed all strike the piston at once and stop dead. The heat energy of the steam would be transferred in its entirety to the piston, and the laws of thermodynamics would, in this one instance, be violated. Such an occurrence is fantastically improbable—as improbable as the proverbial chimpanzee tapping out *Hamlet* at the typewriter—but it is not a strict logical impossibility. In this respect, the statistical view of thermodynamic laws, enunciated by Boltzmann and Gibbs, was something subtly new to physics. In any practical, realistic sense, the laws of thermodynamics are exact and inviolable, as laws of physics ought to be, but philosophically they are not quite one-hundred-percent solid and true.

The success of statistical mechanics was thus a slightly ambiguous one. In a general way, it represented the achievement of a long-standing desire—the ability to understand the gross properties and behavior of objects in terms of the detailed activities of their fundamental constituents. René Descartes had wanted to picture all of nature, human beings included, as machinery of vast and complicated design, and imagined that scientists would eventually understand nature as they understood the workings of a clock. Descartes thought of himself as a considerable physicist, and was forever inventing mechanical models to explain gravity or digestion or climate in terms of wheels, ropes, and pulleys. His science is largely forgotten, but his philosophy took hold. Pierre-Simon de Laplace, a prodigious French mathematician and astronomer who derived from basic Newtonian physics a sophisticated body of mathematical methods for dealing with the behavior of complicated mechanical systems, imagined "an intellect which at a given instant knew all the forces acting in nature, and the position of all the things of which the world consists" and supposed that such an intellect, by application of the known laws of physics, would be able to predict what this machine of nature would be doing at any time in the future, and to say with certainty what it had been doing at any time in the past.[3] This intellect would acquire thereby a perspective of nature that embraced all of time in one glance; in Laplace's words, "Nothing would be uncertain for it, and the future, like the past, would be present to its eyes."

Knowledge of physics, of the rules of nature, would bestow a Godlike view of the world. Complete knowledge of the state of nature at any one time, coupled with complete knowledge of the laws that govern it, is equivalent, in Laplace's argument, to complete knowledge of the state of nature at all times, from infinite past to infinite present.

But statistical mechanics was not as powerful as Laplace would have liked; it is based on knowledge not of the exact position and speed of every atom but merely on the average position and speed of typical atoms. This is enough to allow broad statements about what a volume of gas as a whole will do. The power of statistical mechanics is that it allows the large-scale laws of physics governing macroscopic systems to be obtained without the complete knowledge that Laplace thought necessary. But that power is achieved at the expense of a small loss of certainty. The laws of statistical mechanics are true only in a gross, aggregate way, and they can be broken, in principle, by some perverse but not forbidden accident in the arrangement of a collection of atoms. Laplace's idea that a scientist could understand nature as an unerring machine, as finely predictable as the scientist's calculation would allow, remains in practice unfulfilled.

Statistical mechanics and the atomic view of nature are now so deeply ingrained in our thinking that it is hard to appreciate how controversial they once were. In our modern view, the success of statistical mechanics lends credibility to the atomic hypothesis on which it is founded, but to many physicists of the nineteenth century such reasoning seemed backward. Atoms could not be seen or individually manipulated, and it therefore seemed to these critics that all the successes of statistical-mechanical theory did no more than establish the atomic hypothesis as a useful device—a handy set of mathematical formulas, a convenient abstraction. Statistical mechanics allowed an explanation, to be sure, of certain physical results and ideas, but this was clearly not enough to prove that atoms really existed.

The objections of the anti-atomists were more than fastidiousness; physics had always been counted as a science that dealt with tangible and measurable realities, like the motion of a projectile or the changes in temperature of a steam engine going through its cycle. The bulk properties of matter—heat, energy, mass, electric charge—were directly accessible to the experimenter's hand, and it seemed to many in the nineteenth century that it was with such bricks that physical theory ought to be built. The evidence in favor of the atomic hypothesis was by contrast circumstantial. It was admitted at the outset that atoms themselves could not directly be picked up, isolated, measured, pushed around, or tweezed apart, and that belief in the existence of atoms could not be based on these traditional criteria. Instead, the evidence that atoms were real, and not useful mathematical contrivances, was that a theory based on atoms produced a wealth of satisfactory understanding and quantitative explanations of physical phenomena. The power of the atomic theory as a way of understanding physics at a deeper level appears undeniable today, but in the nineteenth century it was a wholesale departure in the style of physics to reason from unprovable hypotheses. The time-honored method in physics was to observe and measure those quantities that could be directly grasped by experimental methods, and to construct theories linking those quantities to each other. To go beyond this, to propose that at some deeper level there were underlying physical laws and entities whose effects could be seen only indirectly, seemed to some scientists dangerous, even sinful.

This distaste for abstraction hindered acceptance of another achievement now regarded as one of the great intellectual triumphs of the nineteenth century, James Clerk Maxwell's theory of electromagnetism. In the early 1800s, there was a great mass of empirical evidence and practical knowledge of electricity and magnetism, but it was haphazard. There were obvious connections between the two subjects—it was well known, for example, that coils of wire carrying electrical currents created magnetic fields—but nothing like a complete explanation of the many observed phenomena was at hand. As had been the case with heat, there was fierce debate as to the nature of electricity. Positive and negative forms of electricity had long been recognized, but it was unknown whether they were two separate kinds of electricity, or whether one represented the presence and the other the absence of a single electric fluid or substance. Attempts to

construct theories of electricity and magnetism generally began
with the idea that electricity must be not just a material substance, able to cling to objects as static charges and flow through
wires as a current, but also some sort of emanation, allowing
charges and currents some distance apart to influence each other
and to disturb magnetic compasses. If electricity was a substance,
it had to be, like the old heat fluid, an insubstantial kind of substance, able to pass through the body of conducting metals and
spread through the vacuum of space.

The first person to move away from these notions was the great
experimental physicist Michael Faraday, a man of profound physical insight but little traditional learning. He was the son of a
blacksmith, belonged to the newly mobile entrepreneurial class of
industrial Britain, and unlike most of his scientific contemporaries, was a man of few cultural leanings. He was more or less
self-educated in physics, having taught himself first at home, from
books, and then as an apprentice in the London laboratory of the
celebrated chemist Sir Humphry Davy. Faraday was knowledge-
able in physics but not weighed down by the trappings of physical
philosophy or traditional ideas, and was thus free to theorize
about the nature of electricity in whatever way his imagination
led him. Among his many experimental accomplishments was the
long-sought demonstration, in 1831, that magnetic fields could
create electric currents. Because a constant current flowing in a
coil will produce a magnetic field, the natural assumption was
that a constant magnetic field, suitably arranged, ought to be able
to generate electricity, but numerous physicists had tried and
failed to demonstrate the connection. Faraday, however, through
a combination of ingenuity and acute powers of observation,
made the essential discovery that to create a current the magnetic
field must be changing. A magnet placed next to a wire would do
nothing, but a magnet passed over a wire would create a transient
pulse of electricity.

To aid his thinking, Faraday conjured up a pictorial image of
what he called "lines of force," originating in a magnet and running off into space. Near the magnet, where the lines were close
together, the force would be strong; as the lines spread into space
and away from each other, the magnet's potency diminished. By

the same means, Faraday could also make a picture of the magnetic generation of currents. As a moving wire cut through lines of force, or as moving lines of force passed through a wire, a current would be generated whose strength was directly related to the rate at which wire and magnetic lines of force were intersecting.

With these innovations, Faraday set the practical use of electricity and magnetism on its modern course. From the principles he discovered, it was possible to design dynamos and motors to turn mechanical motion into electricity and back into motion again. But although Faraday's practical work was of immediate utility, his theoretical innovations were not much noticed. Lord Kelvin was aware of Faraday's lines of force, but he, the perfect product of a traditional scientific education, evidently found the concept too avant-garde for his taste. It was left to James Clerk Maxwell, another lone genius, to turn Faraday's pictures and lines into the modern theory of electromagnetism. Maxwell grew up on a minor Scottish estate; his talents were recognized early, but his accent caused amusement when, as an undergraduate, he first went to Cambridge. He built himself a solid scientific career, making profound theoretical innovations in atomic theory and celestial mechanics (he argued, for reasons of gravitational and mechanical stability, that Saturn's rings must be made of small particles), but moved from Aberdeen to London to Cambridge, retreating at times to his family lands, and thus maintained a small distance between himself and the scientific community.

Maxwell was from the outset impressed with Faraday's powers of experimentation and imagination. Faraday had acquired, in his self-guided way, an iconoclastic but profound understanding of electromagnetic phenomena, and his ideas set Maxwell on the path toward a new theory of electromagnetism, iconoclastic in its own way. Abandoning the idea of little wheels and cogs transmitting physical influences throughout space, which seemed to traditionalists like Kelvin the right way to make models of electromagnetism, Maxwell turned Faraday's lines of force into a new mathematical construction, the electromagnetic field. At every point in space, he erected a little arrow, of variable length and direction, and he defined mathematical operations by which the local effect of electric and magnetic forces could be found from the disposition of these arrows. He wrote down four equations— Maxwell's equations of the electromagnetic field—that related the length and direction of the arrows to the placement of electric charges and the flow of electric currents. These mathematically defined arrows became known as the electromagnetic field; unobservable in itself, it was derived from charges and currents, and in turn determined what electrical and magnetic effects would be observed.

Maxwell's theory was not wholly an innovation. Kelvin and many others had attempted unified theories of electricity and magnetism, in which there was pictured some kind of fundamental electromagnetic "fluid" with certain physical properties of compressibility and viscosity, and from which electrical and magnetic effects could be calculated (electrical influences might be due to the velocity of this fluid, while magnetic effects were related to its rotation, for example). Such constructions were based on the idea that a true physical model had to be something tangible and reducible to everyday terms; if electricity and magnetism could somehow be analogized by means of a fluid with familiar fluid properties, then electromagnetism would be understood. But Maxwell did away with these models. In his derivation of the electromagnetic field and its controlling equations, he had made use of mechanical analogies and mental pictures of space filled with whirling vortices transmitting their rotation to one another by means of little wheels. But when, in 1864, he published this theory in its modern form, it had been, as one commentator put it, "stripped of the scaffolding by aid of which it had first been erected."[4] The electromagnetic field stood by itself, defined completely by mathematical laws.

Maxwell's theory caught on, but slowly. What we now see as an elegant theoretical structure was seen by many of Maxwell's contemporaries as, like the atomic hypothesis, an inexplicable mathematical abstraction, unsatisfying precisely because it was based on no underlying "physical" model. Kelvin spent many years trying to "explain" Maxwell's theory by deriving it from some fluid or mechanical model of space, as if one could explain a painting by scraping off the colors to reveal the numbers underneath by which the artist had been guided.

What aided the acceptance of Maxwell's theory was that it made some testable predictions. Notably, the theory indicated that an electromagnetic disturbance created at one point, by, for instance, a jiggling electric charge, should propagate from its source and be detectable elsewhere. Because of mostly technological limitations, it was not until 1887 that the German physicist Heinrich Hertz succeeded in passing an electromagnetic signal from a generator to a detector a few feet away, but soon afterward, in 1894, Guglielmo Marconi turned Hertz's laboratory trick into a system for transmitting radio signals across great distances. In 1901, he sent the first "wireless" transmission across the Atlantic Ocean, from Cornwall to Newfoundland. If Maxwell's theory could engender the new technology of radio transmission, it could hardly be called an abstraction.

There was intellectual achievement here, too. In his equations, Maxwell had included two numbers, experimentally derived, that specified the strength of the electric and magnetic forces, and from these two numbers the speed at which electromagnetic signals should travel could be calculated. It turned out, unexpectedly, to be the speed of light. This was a surprise and a triumph: two numbers obtained from the static properties of electric charges and magnets showed up, in thin disguise, as a third fundamental constant of nature. As Hertz and others continued their experiments on the transmission of electric signals, it dawned on them that light itself was, in fact, an electromagnetic phenomenon; this realization was one of the great intellectual leaps in scientific history. Maxwell's theory was not just a unified theory of electricity and magnetism but a theory of light as well.

Late in the nineteenth century, physics appeared to be fast approaching a kind of perfection. The theories of mechanics and thermodynamics, themselves unified by the ability of statistical mechanics to explain heat and energy in terms of atomic motion, seemed able in principle to say everything that needed to be said about the nature of matter. Maxwell's electromagnetic equations linked disparate laws and phenomena collected over the previous couple of centuries and, as a quite unwarranted bonus, described light too. Standing somewhat alone, but perfect in itself, was Newton's inverse-square law of gravity. By no means was every physical phenomenon explained and accounted for, but it is easy to see why physicists of that time began to think that all the explanations were theirs for the asking. They had a theory of everything they thought they needed a theory of, and a total comprehension of the physical world would be had, apparently, as soon as the ingenuity of physicists allowed. Fundamental physics seemed to be nearing its end, and all that would remain would be the technically daunting but perhaps intellectually unsatisfying task of generating from first principles detailed explanations of complex macroscopic phenomena. A great intellectual endeavor, begun by the ancient Greeks and set on its modern path by Galileo and Newton, was pretty much wrapped up.

With the hindsight of a hundred years, we can see what went wrong with this, and we know which nigglingly inexplicable experimental results were the ones that classical physics in all its glory would fail to account for. But even at the time there were evident problems, although one might have thought of them as philosophical more than scientific difficulties. Most notably, there was a certain antithesis between Maxwell's theory, which was founded on a smooth, continuous, infinitely variable quantity called the electromagnetic field, and atomic physics, which was based on the existence of discrete, finite, indivisible units of matter. Doubters of both theories worried at the increasing reliance on indirect mathematizing to explain what were, after all, supposed to be physical phenomena of the real world. Opponents of atomism found atomic theory objectionable because it based explanations on the behavior of invisible microscopic entities; critics of Maxwell's theory regarded the electromagnetic field as a mathematical abstraction, undetectable in itself and apparent only through the numerous electromagnetic phenomena it mediated. The practical successes of Hertz and Marconi surely indicated that detectable electromagnetic influences could propagate through empty space, but to many physicists it seemed that electromagnetic transmissions according to Maxwell's theory were mysteriously disembodied; if radio signals moving through space were thought of as some sort of electromagnetic equivalent of ocean waves rippling across the surface of the sea, what, in Maxwell's theory, represented the sea? What was the medium upon which electromagnetic waves traveled? According to Maxwell, disturbances in the electromagnetic field existed by themselves, devoid, as far as anyone could tell, of physical support. This struck many traditional physicists, such as Kelvin, as self-evidently insufficient, and a great deal of ingenuity was spent in thinking up models, to be attached to Maxwell's theory, for a medium in which electromagnetic waves could travel. This medium was called the "luminiferous aether."

The ether, to use the conveniently shortened modern name, was a mechanical model of space, and its inventors used endless ingenuity to devise various versions. One might start with rods connected to balls, but this construction would be rigid and would not allow the passage of waves, so the rods had to have hingelike connections to facilitate movement—but then the whole structure would not be rigid enough, and would transmit too many kinds of waves, so to each rod a gyroscope was attached, fixing its direction in space unless some additional force was supplied. The attitude of nineteenth-century physicists to these museum-piece theories is hard to guess. They did not think that space was literally filled with rods and gyroscopes, but they did believe that until the properties of space were explained according to the established Newtonian principles of mechanics, physics would not truly be understood. It was in these attempts to set electromagnetism on a mechanical foundation, however, that one of the two great flaws of classical physics showed up.

Any mechanical ether was, by definition, a fixed medium through which electromagnetic phenomena such as light or radio waves traveled at some constant speed. But the Earth is also traveling through space (orbiting around the Sun at about 30 kilometers per second, or 0.01 percent of the speed of light), so the speed of light as measured by experimenters on Earth ought to be slightly different depending on how fast the experimental device, sitting on the Earth's surface, happens to be traveling through the ether. This necessary consequence of all ether theories, regardless of the details of their construction, was put to the test in a famous experiment conducted by the Americans Albert Michelson and Edward Morley in 1887. In their apparatus, a beam of light was split into two beams perpendicular to each other, which were sent out a short distance to mirrors and reflected back to the center of the device. They were then able to look for the difference in the time taken for the two beams to traverse their perpendicular paths, and the apparatus could be rotated to see whether the times varied according to the orientation of the beams with respect to the Earth's motion through the ether. If the ether was real, variations in time should have been easily detectable, but Michelson and Morley found nothing; there was no discernible change in the speed of light no matter how the experiment was positioned.

The negative result of the Michelson-Morley experiment was a complete surprise. What should have been a routine confirmation of a standard piece of physical theory was instead a direct contradiction of it. The negative result could be taken to mean that the Earth was perfectly at rest in the ether, but this was wholly unreasonable: the Earth travels around the Sun, and it would be bizarre if the ether, which supposedly filled the entire universe, should share the motion of our planet among all celestial objects.

One way of fixing up the problem was to grant that the Earth was indeed moving through the ether at some unknown rate but to suppose also that the length of objects moving through the ether changed just enough to compensate for the change in the speed of light. Light would then travel a little more slowly through one arm of the Michelson-Morley apparatus, but that arm would be reduced in length by an exactly equivalent amount, so the time needed to make the round-trip would remain precisely the same. This suggestion seems to have been first made by the Irish physicist George FitzGerald, but it was quickly taken up and developed into a full-scale theory by many others, particularly the Dutch physicist H. A. Lorentz and the French mathematician Henri Poincaré. Poincaré in fact formulated a new principle of physics, which he called the principle of relativity: it asserted that no experiment of any kind could detect an observer's motion through the ether. Applied to the idea of contraction proposed by FitzGerald, Poincaré's new rule dictated that in all cases lengths must change exactly to compensate for changes in the speed of light, and this requirement alone was enough to dictate the mathematical form of the theories worked out by Poincaré and Lorentz. The day was saved, so they hoped.

But the ether was by now a strikingly odd contrivance. At bottom, it was still a mechanical construction, obedient to the laws of Newtonian physics and therefore possessing some absolute state of rest. But since Poincaré had declared this state of rest to be undetectable, the interaction of matter with the ether had to have the peculiar property of altering the lengths of measuring sticks in such a way that any experiment designed to detect motion through the ether would be defeated and give a result indistinguishable from what would occur if the observer were truly at rest in the ether. This last hurrah of the mechanical ether models in the early years of the twentieth century was thus a wholly unsatisfactory arrangement: the urge to preserve the stationary mechanical ether was so strong that an unexplained interaction between ether and matter was dreamed up with the sole purpose of thwarting any experimental attempt to detect it.

The real problem, as Poincaré in particular began to understand, was that mechanics according to Newton worked in one way, and the "ether," if there was such a thing, had to work in another way in order to meet the requirements of Maxwell's theory of electromagnetism and light. The dynamics of the electromagnetic field and the dynamics of solid bodies did not appear to obey the same rules, and this was why mechanical models of ether were bound to fail, unless one was willing to introduce some arbitrary extra ingredient such as the FitzGerald contraction.

This was not the only point at which the elements of classical physics seemed to have some difficulty in meshing neatly together. The second great flaw in the edifice showed up when physicists pondered a simple question: Why does a poker placed in a fire glow red? A hot poker, according to the atomic theory, was hot because its atoms were jostling fiercely; a red glow, physicists now knew from Maxwell's theory, was due to the emission of electromagnetic radiation of a certain characteristic wavelength. (Newton had been the first to prove experimentally that what we see as white light is a mixture of all the colors of the rainbow; by now it was understood that each color of light corresponds to a certain band of electromagnetic radiation, red being of longer wavelength than blue.) If classical physics were to pass the seemingly simple test of explaining the red glow of a hot poker, it would have to explain why the jostling motion of the atoms turned into radiation, and why that radiation was of a particular wavelength.

A preliminary understanding came easily and quickly. The ingenious thermodynamicists of the nineteenth century deduced that just as the heat of the poker is characterized by its aggregate atomic motions, so the energy of electromagnetic radiation could be characterized as the sum of all the energies of individual electromagnetic waves within it. The heat of the poker and the electromagnetic energy it emitted had to be related by a thermodynamic balance, just as there was a balance between the heat of atoms in a steam engine and the mechanical energy transmitted to the piston. From general principles it was proved that the radiation emitted by a body—its color and intensity—must depend on the temperature of that body and on nothing else.

This was obviously a move in the right direction. It was known experimentally that a heated body begins to glow red, emitting radiation at the long-wavelength end of the spectrum, then as it is heated glows whiter, as different colors are mixed in, and if it can be made hot enough finally turns bluish, as light of yet shorter wavelengths is produced. The thermodynamic argument indicated why the color changed as the temperature rose, and what remained to be worked out was exactly how the color of the radiation was related to the temperature.

This, sadly, was where things went wrong. From thermodynamics, it was deduced that a body at a certain temperature ought to emit energy at a certain rate: the hotter the body, the more radiation it would put out. But when physicists tried to calculate from elementary principles the composition of that radiation—the quantity of energy emitted at each wavelength—the answer came out disastrously wrong. In all the calculations, the jostling of atoms in a hot body invariably gave rise to more radiation at short wavelengths than at long ones. In other words, even a warm poker would, according to the theory, produce more yellow and blue light than red light, and would therefore glow blue-white, not red. Classical physics, in the attempt to marry thermodynamics with Maxwell's electromagnetic radiation, failed to explain the red-hot poker.

In this case, as with the problem of the mechanical ether, it seemed that pieces of the overall structure of classical physics rested on different—and incompatible—foundations. Although physics in general was forging ahead, with all kinds of problems in thermodynamics or electromagnetism being posed and solved, the points of contact between parts of the discipline revealed contradictions where there should have been mutual reinforcement. The much vaunted "end of physics" that almost came to pass in the nineteenth century seems in retrospect mostly an illusion: as the superstructure was growing, cracks were opening in the ground.

Lord Kelvin, perhaps fortunately for his reputation, was too old by the turn of the century to pay much attention to these difficulties. Fifty years earlier, he had made use of the certainties of classical physics to deny the geologists' and biologists' demands for an older Earth than thermodynamics would allow, and had mocked their aspirations to be real scientists—analysts and theorizers rather than mere classifiers and catalogers. But he lived long enough to see his own certainties attacked. In 1896, the French physicist Henri Becquerel reported invisible emanations coming from certain naturally occurring minerals; these emanations fogged photographic plates sealed in black paper, but could not be otherwise detected. In 1903, Pierre Curie discovered that certain radium compounds produced not only invisible emanations but heat too. On top of the theoretical failures of classical physics (the problematic ether, the inexplicable red-hot poker) now arrived strange new phenomena—experimental results that had not been predicted at all, and which had, for the time being, no place in the known world of theoretical physics. The discovery of radioactivity began a new chapter in physics; the elucidation of radioactivity became the chief work of the opening decades of the twentieth century. Talk of the end of physics, if there had ever been any, was heard no more.

But before anything was really understood of the nature of radioactivity, a simple fact was grasped. If ordinary rocks could emit some new kind of energy, then Kelvin's argument that the Earth could be no older than a hundred million years was overthrown, for it was an essential part of his argument that the Earth's interior was an inert, cooling body, born at a certain temperature and inexorably losing heat. But put in a fresh source of heat, able to keep the Earth gently warm even while its original heat leaked slowly into space, and it could be much older than Kelvin had declared.

In the absence of any appreciation of what radioactive heat really was, there was now no way to put any kind of limit on the age of the Earth. It might just as well be infinite—although the geologists had by this time so accustomed themselves to straitjacketing their theories into a hundred million years or so that they resisted attempts to make the Earth very much older. It fell to Ernest Rutherford, the boisterous New Zealander who dominated the new science of nuclear physics well into its adolescence, to summarize, during a lecture at the Royal Institution of London, the significance of radioactivity for the debate over the age of the Earth. The elderly Kelvin, who was in the audience, slept through most of the talk, but, in Rutherford's famous account of the evening, "as I came to the important point, I saw the old bird sit up, open an eye, and cock a baleful glance at me! Then a sudden inspiration came to me, and I said Lord Kelvin had limited the age of the Earth, *provided no new source of heat was discovered.* That prophetic utterance refers to what we are now considering tonight, radium! Behold! The old boy beamed upon me."[5]

Kelvin in fact never conceded that radioactivity would allow the Earth to be much older than he had originally said. Since no one then knew where the heat was coming from or how much there might be, he was at liberty to believe that radium was ultimately inconsequential, although he backed up his assertion with an old-fashioned and implausible idea that radium somehow extracted energy from the ether and redistributed it as heat. He died in 1907, having taken ever less notice of radium. Rutherford and everyone else realized, however, that the vexed question of the Earth's antiquity, on which Kelvin had spent so much of his energy, was now completely up in the air again. It might seem that the physicists would have had to admit that they had been wrong and the geologists and biologists right. But they were for the most part unabashed, and Rutherford's inspired apologia for Kelvin provided a way for them to think that they had been correct in principle and wrong only in detail. They were right to say that the age of the Earth must be limited by physics, and they were mistaken only in that they did not then know of certain undiscovered phenomena that would invalidate the particular numbers they came up with. But for not taking into account what had not been discovered they could hardly be blamed. Physicists still ought to aspire to certainty, though the facts on which that certainty rested might change; physics still should aim at the infallible pronouncement.

2

Beyond Common Sense

CLASSICAL PHYSICS at its zenith had an influence that spread across all of science, and beyond. Geologists and biologists, forced by Lord Kelvin's strictures on the age of the Earth to come up with a properly accounted history in place of a vague, uncalculated past, adopted a more rigorous scientific method, which they maintained even after Kelvin was proved wrong. The manner of physics, with its theories, mathematical models, and quantitative reckonings, became the style to aim for in every discipline of science.

The seeming comprehensiveness of physics appealed to thinkers in other disciplines, too. Historians traditionally thought of their subject in part as the elucidation of cause and effect (Oliver Cromwell beheaded Charles I as a final consequence of popular discontent, which was due to Charles' high-handed treatment of parliament, which derived from the King's excessive belief in his divine prerogative to rule, which he acquired from . . .) and so were susceptible to the idea of casting history as, so to speak, a version of statistical mechanics. Where the physicists saw order emerging from the aggregate effects of individually unpredictable atomic motions, so the historian wanted to see historical evolution as the gross consequence of individual human events. In the later chapters of *The Education of Henry Adams*, the author makes strenuous and endearing attempts to portray the vast changes in society at the end of the nineteenth century as mechanistic "accelerations" caused by historical "forces." Karl Marx, especially, wanted to make history "scientific," and talked of forces that would push the development of societies in certain ways, away from the past and into the future. Both Adams and Marx borrowed from biology the idea of historical evolution, of adaptation and progression. But the attempt to bring scientific order to the social studies was largely totemic, as if the historians and social scientists, by adopting the language of the physicists, could partake of the certainty and intellectual rigor of physics. History portrayed in terms of large-scale forces, a history of actions and reactions, sounds more reassuringly explicable, more rational, than history seen as a series of events dictated by the unpredictable personalities and emotions of a handful of powerful figures.

But there are no fixed laws governing historical causes and effects as Newton's laws of motion govern masses. Marx and others transformed the study of history, but it is hard to believe that their ideas were in any deep way a consequence of the science of the late nineteenth century, no matter how much they might have appropriated scientific language or aspired to scientific ideals: this was a time of change in all things, and historians were part of the crowd, breathing the same atmosphere that physicists, geologists, painters, and poets breathed.

The reaching of physics toward perfection had direct and far-reaching consequences in philosophy and theology, however. The Marquis de Laplace had asserted that if we understood in perfect detail the mechanical disposition and workings of everything in the world at one instant, then by extrapolating forward and backward in time, we would be able to see the entire past and future of the world laid out before our eyes. We would view history as a narrative tapestry; endowed with complete knowledge of physics, we would have acquired part of what St. Augustine had decided was the essence of God: an intellectual vantage point outside time, from which all of time could be seen in one encompassing gaze.

Classical physicists of the nineteenth century never reached, in any practical sense, the degree of knowledge they would have needed to fulfill Laplace's ambition, but the seemingly inescapable conclusion that they would one day do so raised all manner of questions about determinism and free will. Laplace's assertion meant, for example, that the atoms that came together in 1756 to form Mozart's brain, and which in the course of thirty-five years turned out apparently original music, did so in an entirely predictable and predetermined way; even when atoms first came together to form the Solar System, they contained in their motions and interactions the seeds of everything that followed. This raised difficulties for understanding free will, and in a paradoxical way began to make nonsense of physics, too. We conventionally learn that Mozart's music was a product not only of his own creative abilities but of his knowledge of the work of earlier composers: his father, Bach, Haydn. But if Mozart's brain and symphonies were already programmed in the dust that formed the Earth, it made no sense to imagine that his music in any way derived from its antecedents. The dust that formed the Earth must have contained, all at once, the ingredients not just for Mozart but for Mozart's father and for Bach and Haydn. In the godly perspective Laplace accorded physics, Mozart and his forebears are coevals instead of succeeding actors in a developing drama.

The idea of cause and effect, that one action precedes and is essential to the next, is fundamental not only to musicology but to physics. By insisting on the supremacy of mechanical cause and effect at the level of atoms, and thus concluding that the whole of history is already contained in the pre-planetary dust of the unformed Solar System, physicists appeared to be saying that cause and effect, in the macroscopic sense and at later times, means nothing. The viewpoint forced by physics has, ironically, led back to a philosophical contrivance of the Renaissance, whose theologians were able to reconcile God's foreknowledge of the world with the apparent ability of human beings to exercise free will. We think, for example, that when we raise a glass in hand in order to drink, we first decide we want a drink, then send the appropriate instructions to the muscles of our arm and hand to perform the task we have chosen. This is free will, but the medieval theologian says it is an illusion: when God created the world He had already built into it a later moment when I would decide I wanted a drink and would therefore raise a glass to my lips. So careful is God's engineering that, in an entirely predetermined manner, the notion of being thirsty strikes my brain the moment before the muscles of my arm begin to move. I think I have made a decision and acted upon it, but in reality both the thought and the action were laid down in advance.

Fanciful as this argument might seem, the eighteenth-century philosopher David Hume could find no logical way to dismiss it, and therefore came to believe that we could never distinguish with certainty between any apparent instance of cause and effect, and mere coincidence. Scientists build their theories on the assumption that cause and effect is real, and that one thing happening after another betokens mechanical connection. But the perfection of classical physics, were it to be achieved, would have made Hume right, and cause and effect meaningless.

These questions are real, deep, and unresolved. Perhaps it is fortunate that as they loomed physics began to lose its aura of perfectibility, and Laplace's ideal of complete knowledge and complete predictability began to recede. Physicists had come across problems in classical physics, notably their failure to understand radiation from hot bodies and their inability to construct mechanical models of the ether, and when physicists have quantitative matters to resolve, philosophy seems an especially useless occupation. At the turn of the century, these two niggling problems had grown to threaten the whole structure of classical physics. Physicists were in the position of archaeologists who think that they have dug down to the earliest foundations of Troy, and then find a couple of bricks from the still-buried walls of a whole new city. What was thought to have been original becomes merely an outgrowth from an earlier time. So it was in physics. The difficulties over radiation theory and the mechanical ether exposed fundamental incompatibilities between the classical theories of mechanics, radiation, and atoms. The cracks ran through the foundation and lower, and the innovations that explained these difficulties— quantum mechanics and the theory of relativity—were indeed attached to a wholly new foundation, below the ground level of classical physics.

Some of the language of quantum mechanics and relativity, like that of classical physics, has passed into common usage. Quantum theory in common parlance means the uncertainty

principle, which is taken to mean that even physicists can never
be quite sure of what they are measuring; this is perceived to
have something to do with the quantum-mechanical fact that
every act of measurement affects the thing measured, so that
there can be no objective truth. Relativity is taken to mean the
inability of people to agree on the length of measuring rods or
the time shown by their clocks, which seems to mean that two
people can have different views of the same matter yet be
equally right; this, too, suggests that there is no such thing as
objective truth. Modern physics, in other words, is popularly
understood to have done away with the essential classical ideal
of an objective world existing securely out there, waiting to be
measured and analyzed by us. Instead, we must think of ourselves as inalienable participants of the world we are trying to
understand, inextricably tied to the things we are trying to comprehend. The possibility that we can understand all of physics
and thus all of the world becomes a logically unattainable goal,
because we ourselves are in the world, affecting it at the same
time that we are trying to stand outside it and apprehend it.
Determinism is blown down, free will restored, history given
back its arbitrariness and humanity its spontaneity.

All this, comforting as it might sound to anyone worried by the
overwhelming power of classical physics, is a garbled version of
the truth. Quantum mechanics and relativity indeed change some
of the fundamental notions on which physics is built, but it is a
gross misunderstanding to imagine that they introduce into science some kind of arbitrary subjectivity. It is true that under the
new law one has to take account of the measurer in order to
define and understand the thing measured, but there are rules for
doing this accounting; it is not whimsical.

Albert Einstein, who devised the theory of relativity, would have
been able to communicate with Isaac Newton: they spoke the
same language—or, rather, Einstein spoke the language that Newton had largely invented some two hundred years before. Both
Einstein's relativity and Newton's mechanics are theories that deal
with the motion of bodies, idealized mathematically as point
masses moving along precisely defined lines and subjected to

exactly measurable forces. The differences lie in the rules used to calculate the trajectories of bodies and the response of the bodies to forces.

Historical accounts of the genesis of relativity often start with the negative result of the Michelson-Morley experiment: mechanical ether models required the speed of light to be different for observers moving in different directions through the ether, but the Michelson-Morley experiment showed that the motion of the Earth through the ether had no discernible effect on the speed of light. Einstein, however, always maintained that he had not heard of the Michelson-Morley result when he struck upon the special theory of relativity, and that he devised the theory by pure force of mind, on consideration solely of the logical consistency required of any complete theory of motion. His point of departure was a theoretical dilemma. Maxwell's electromagnetic theory allowed the theoretical physicist to perform a purely mental version of the Michelson-Morley experiment: one could construct a mathematical description of light traveling through space, and of it inquire how light would appear to observers moving at different speeds. One of the straightforward predictions was that someone traveling toward a source of light should see it shifted to higher frequency, so that red light would become bluish. This corresponds to the Doppler effect for sound, well known at that time, by which the siren of an ambulance approaching a bystander sounds higher in pitch and turns lower after the ambulance goes past. But the same mathematical description of light also yielded a surprising, seemingly nonsensical result: although the wavelength of the light would change for different observers, its speed would not. This seems to be an evident impossibility. If we are moving toward a source of light, then surely that light must strike us with a higher speed than if we were stationary or moving away. How can it be otherwise?

Yet the mathematics of Maxwell's equations insists that the speed of light is always the same, regardless of one's state of motion, and the negative result of the Michelson-Morley experiment backs up this assertion. Einstein may or may not have known about Michelson-Morley, but he surely knew about the mathematical problem. Henri Poincaré was the first to see clearly what was wrong with classical ether models: it was not that physicists had been insufficiently ingenious in finding an ether that worked but that all ether models relied on Newtonian mechanics, and Newtonian mechanics could not accommodate the properties of light dictated by Maxwell's theory of electromagnetism. In Newton's world, all speeds, whether of solid bodies or of light, were measured against the background of an absolute space, and had necessarily to be different for observers who were themselves moving, but according to Maxwell light did not obey this seemingly inescapable law.

This was the conundrum that Einstein, at that time working privately at physics in the spare time from his job as a patent examiner in Bern, set himself to solve. His approach to it, characteristic of his attitude toward all the physical problems he addressed in his early years, might be called the Sherlock Holmes method: attend strictly to the facts, and when all the likely explanations have been dismissed what remains, no matter how unlikely, must be the truth. If Maxwell's theory predicts that light always moves at the same speed, it was no good trying, as FitzGerald and Poincaré had tried, to pretend that light really behaves according to Newtonian rules, that there really is a mechanical ether, and that measuring sticks and other bodies moving through the ether change their lengths in just such a way as to make it appear that the speed of light is constant even though "underneath" it is not. Instead, Einstein had the courage to accept the demonstration, from Maxwell's theory, that the speed of light was constant, and he set out to discover what this constancy implied.

If one of the marks of genius is the ability to look at obvious facts and discern that they are based on concealed and undiscussed assumptions, then Einstein was truly a genius. To the physicists of the first years of the twentieth century, it made no sense to countenance Einstein's idea that the speed of light could actually be constant. Imagine two boatmen, one approaching a lighthouse, one drawing away from it, and imagine that just as they pass each other they both see a flash of light from the lighthouse. One boatman is moving toward the source of light and the other is moving away; it seems inescapable that they will measure the speed of light differently. Let us say that the boatmen know that the lighthouse emitted its flash of light at midnight; the flash would reach them a fraction of a second afterward, having traversed the distance from the lighthouse to their boats. Knowing the distance to the lighthouse and the time at which the flash reached them, the boatmen could figure out the speed of light. The boatman approaching the lighthouse would presume that the measured speed of light would be the true speed of light plus his own speed, and the receding boatman would measure a speed less than the true speed of light. This seems impossible to refute.

But Einstein raised a practical question. How, in this example, do the boatmen know for sure that midnight for the lighthouse keeper is the same thing that both the boatmen know as midnight? All three might have synchronized their clocks some time ago, but that would have been at a different time. What would the boatmen do if they wanted to be sure that all three clocks were still synchronized? The elementary answer, as Einstein observed, is that they would have to communicate, but if they want to communicate they have to exchange signals—for example, by sending flashes of light to one another. And here is the problem: in order to be sure that they are all measuring the same thing, the boatmen want to have some way of being certain that the flash of light that the lighthouse keeper sent out at midnight by the lighthouse clock was sent out at midnight by the boatmen's watches; but the only way they can verify this would be by having the lighthouse keeper send out by pre-arrangement a light flash when the lighthouse clock strikes midnight—which is precisely what they are trying to check.

Einstein thus identified a subtle weakness in the Newtonian picture. It is assumed that all clocks strike midnight at the same moment, and on this assumption one is forced to conclude that the speed of light would have to be different for the two boatmen. But there is no way to be sure that the lighthouse clock and the boatmen's watches reach midnight simultaneously that does not itself involve the passage of light, or some other signal, between the participants. Einstein proceeded to a new realization: it makes sense, in an empirical, verifiable way, to talk about events happening at the same time only when they also happen at the same place. The lighthouse keeper and the boatmen can all get together to set their watches, but afterward, when the lighthouse keeper is back in the lighthouse and the boatmen are off in their boats, they cannot verify that their clocks and watches remain synchronized. The boatmen may know for a fact that the lighthouse keeper sends out a flash of light when the lighthouse clock strikes midnight, but they can only assume that this happens when their own watches read midnight—and that is an assumption on which their deductions about the speed of light crucially depend.

This turns the question of the constancy of the speed of light into something intimately tied up with the nature of time, and it turns the whole package into something that can be tested by experiment. In our common experience, watches stay synchronized, as nearly as we can tell, no matter how we run about. The commonsense idea is therefore that watches remain synchronized regardless, and this is tantamount to the supposition that time is, as Newton assumed, universal or absolute. But if time is absolute, then—as the example with the lighthouse and the boats showed—the speed of light cannot be the same for everyone. So, being empirically minded scientists, we follow Einstein in taking for a fact the constancy of the speed of light, and then turn our reasoning around to ask what it means for the nature of time if the speed of light is indeed constant regardless of the observer's state of motion. The answer is the special theory of relativity.

In short, Einstein posed a choice of systems: if time is absolute, the speed of light cannot be constant; and if the speed of light is measured (or deduced, from Maxwell's equations) to be constant, time cannot be absolute. This is his fundamental and disquieting conclusion: time is no longer public property, but private. Instead of the absolute time of Newtonian physics, which everyone could agree to, there is only the time that each of us carries around with and reckons from the clock we have in our possession.

The common reaction to this novelty, certainly a common reaction in Einstein's day and frequently the reaction of students of relativity today, is that something must be wrong. The notion of time as a universal and shared standard of reference is deeply ingrained, and the idea that the speed of light could be the same for someone moving toward a lighthouse as it is for someone moving away seems so self-evidently contradictory that the struggle to understand relativity is a struggle against common sense. But common sense, as Einstein himself once remarked, is the set of all prejudices acquired by the age of eighteen. Time is certainly a natural phenomenon (the theoretical physicist John Wheeler is said to have remarked that time is what prevents everything from happening at once) but it is also, as used in theories of physics and described in mathematical language, a human invention. The idea of absolute, universal time, agreed to and used by all, came from Newton not because he did experiments to establish how time flowed but because in order to set out a mathematical, quantitative theory of motions and forces he made the obvious, simple assumption. There was nothing in Newton's physics that proved the existence of absolute time; it followed from no empirical analysis, nor from any more fundamental assumption. It was an axiom, something taken to be true because it seemed right and led to a theory that worked. Thus the notion of absolute time, the idea that time meant exactly the same thing everywhere and for everyone, was built into Newtonian physics, and because Newtonian physics was so successful its version of time came to seem unarguable.

Isaac Newton himself was in fact perspicacious enough to recognize that absolute time was not a demonstrated fact, and among the postulates he carefully set out at the beginning of his treatise on the laws of motion was one that said "Absolute, true, and mathematical time, of itself and from its own nature, flows equably without relation to anything external."[1] The generations of scientists who followed Newton mostly lacked his keenness of mind, and took the existence of absolute time to be so obvious as not to bear remarking on. In this particular sense, Newton would indeed have understood Einstein, who seized on this one unproved element in Newtonian theory and, by altering it to accord with the facts, rebuilt our understanding of space and time. Both Newton and Einstein were keenly aware that what is constructed with assumptions rather than by demonstration is not guaranteed to survive. For most of the rest of us, who grow up with common-sense views derived from our learning and experience, absolute Newtonian spacetime seems to be right as a matter of course, a model not needing investigation or verification. This is why the shift from Newtonian to Einsteinian thinking produces a moment of panic, of feeling that the ground at one's feet has turned into air, with the true solid ground far below. What happened to absolute time? Where can it have gone?

But for those striving to understand why two observers can no longer agree on what they mean by time, the only real retort is "Why should they be able to agree?" In our ordinary experience we do not travel at speeds close to the speed of light, and the differences that accrue between clocks in motion are correspondingly tiny. My wristwatch and the clock on the church tower agree to measurable accuracy no matter how wildly I might drive about town. Einstein declares that we should not, and in fact must not, assume that all clocks always read the same, but in disposing of this false assumption he does not leave us floundering. His special theory of relativity is a new set of rules enabling us to understand how the time measured by one person relates to the time measured by another.

The principle of relativity, as first enunciated by Poincaré, was that no one should be able to detect the existence of the supposed luminiferous ether by measuring differences in the speed of light. Einstein transformed this into his own much simpler and entirely empirical principle of relativity, which declared directly that the speed of light is the same for everyone. This made moot any discussion of what the "real" speed of light was or of whether the ether really existed; the ether was a stage prop for classical physics that the new physics no longer required. Einstein banished not only the ether but also any idea of absolute time or space. His principle of relativity stated that for all observers the laws of physics must be the same; since there is no absolute space or time, we can measure and usefully discuss only the relative motion of one observer with respect to another, and we can calculate just as well from either point of view. But physics still has rules, and the rules are the same for everyone. Relativity removes from physics the authoritarian rule of classical physics, with its absolute space and time, and replaces it not with anarchy, in which all participants have their own rules, but with perfect democracy, in which the same rules govern all.

But it is the loss of absolutes, nevertheless, that makes learning relativity such a dislocating experience. To the question of why Newton is wrong and Einstein right in their portrayal of space and time—of why our common sense lets us down—the only solid answer is that our common sense derives from experience in a world where velocities are tiny compared with the speed of light and where the Newtonian depiction of space and time is adequate. But we have no right to suppose that what works for us must automatically be universal. It is a matter of scientific hypothesis and deduction to figure out how space and time behave when bodies are moving close to the speed of light. Naturally, one begins by guessing that the familiar rules will hold, but this turns out not to work, and the rules must be modified in the way that Einstein deduced. It may seem unsatisfactory to respond that there can be only one correct theory of space and time, and that Einstein's happens to be it, but for the physicist such an answer has to suffice. Relativity, like other physical theories, is a set of rules based on a number of crucial assumptions, and experiment and observation bear it out. That is all we ever ask of physical theories, and to ask for some further statement of why the special theory of relativity supplanted absolute Newtonian space-time is to search for a truth beyond the domain of science.

By now, relativity has become a manageable and uncontroversial way of thinking. Physicists who work with it simply get used to it. All the perplexities that strike novices with such unsettling force turn out, in the end, to result from trying to understand the rules of relativity in terms of the rules of classical physics, as if relativity were something merely pasted over the top of the old physics and not a wholly different theory. But once one learns to think exclusively in the language of relativity, the view becomes clear again. Absolute space and time are taken away; the measurement of time becomes personal, not universal; but in other respects things are not much different. The essentials of relativity, as of classical mechanics, are objects idealized as mathematical points moving along mathematical lines. Space and time are still formed from a web of interconnecting trajectories, and the shift from the classical to the relativistic view is largely a matter of understanding that there is no one correct way of labeling the trajectories, that different participants will tend to label and measure things differently. Nevertheless, as long as all the participants agree that one set of rules—those devised by Einstein—is correct, there is no barrier to communication, and, as in the world of classical physics, each participant can in principle have perfect knowledge of the whole arrangement.

The relativistic world is in this sense the classical world seen through a distorting lens. The same ingredients are there, but in a different perspective, and at bottom there is still the notion of an objective world that scientists attempt to grasp in ever increasing detail; the more they discover, the more they will understand the present and the better they will be able to predict the future. Quantum theory, on the other hand, presents an entirely new picture. It puts a frosted screen in front of classical physics, leaving the broad outlines unchanged but erasing detail, so that at the subatomic level there are limits to what the scientist can know. The founding principle of classical physics is that a real, objective world exists, a world the scientist can understand in limitless detail. Quantum theory takes away this certainty, asserting that scientists cannot hope to discover the "real" world in infinite detail, not because there is any limit to their intellectual ingenuity or technical expertise, nor even because there are laws of physics preventing the attainment of perfect knowledge. The basis of quantum theory is more revolutionary yet: it asserts that perfect objective knowledge of the world cannot be had because there is no objective world.

When we try to measure something, we do so with measuring devices that are part of the physical world we are exploring and cannot be disconnected from the things we are trying to measure. No device can measure, for example, the speed of an electron without somehow upsetting the electron's motion, and any measurement of the electron's speed must be qualified with a statement of how the measurement was obtained. Whether in classical physics or in quantum physics, our knowledge of the world is strictly equal to the results of all the measurements we make, but where the classical physicist assumed that measurements could in principle be made both infinitely precise and perfectly unobtrusive, the quantum physicist knows that the thing measured is influenced by the measurement, and will act quite differently when measured than if it had never been measured at all. It makes no good sense to talk of an objective world of real facts if those facts cannot be apprehended without altering them in the process. There is no longer any meaning to be attached to the idea of a real, objective world; what is measured, and therefore known, depends on the nature of the measurement.

This was a violent upheaval in scientists' understanding of what they meant by knowledge of the physical world, and it did not come all at once. The seed for this revolution was the inability of classical physicists at the end of the nineteenth century to understand the apparently simple problem of the radiation emitted by hot objects. The experimental facts seemed simple enough. A body held at some constant temperature emitted radiation—heat and light—over a range of wavelengths and frequencies (wavelength is of course the length of a wave, and frequency is the number of waves per second: as wavelength gets smaller, frequency gets higher), and the bulk of the radiation came out at a frequency that rose in strict proportion to the temperature. At moderate temperature, most of the radiation came out as red light, and as the temperature of the body increased, the emission moved toward yellow and then blue light, which is of shorter wavelength and higher frequency. The spectrum of radiation—that is, the relative intensity of radiation coming out at different frequencies—was a smooth curve, peaking at one frequency, with the intensity falling off for higher and lower frequencies. The shape of the spectrum was found to be essentially the same, regardless of temperature, the whole curve moving bodily to higher frequency as the temperature rose. Simple mathematical formulas had been found for its shape but physicists were at a loss in trying to explain it. In the classical view, adapted from the statistical mechanics of atoms, physicists tried to think of each frequency of light as a basic unit of radiation, capable of carrying energy. Because higher-frequency radiation has shorter wavelength than lower-frequency radiation, any volume of space will be able to accommodate more units of higher-frequency radiation than of lower-frequency. This led directly to the physicists' problem: whenever they tried to calculate, using this picture, the spectrum of radiation from a body at some fixed temperature, more energy always came out at higher frequencies than at lower ones, because more units were available to carry the energy. Instead of peaking at some particular frequency, as experiments showed it must, the theoretical radiation spectrum of classical physics went off the scale, in what came to be called the ultraviolet catastrophe.

One of many people who spent years working on this problem was the German physicist Max Planck, an expert in thermodynamics and statistical mechanics. Like many who puzzled over the radiation spectrum, he fiddled with models hoping to find one that would generate the shape of the spectrum from some straightforward assumptions. In 1900, he found such a model, but its meaning was hard to fathom. As a mathematical trick, Planck decided to add a restriction to classical theory: he specified that the energy carried by radiation could be only a whole number of energy units. His hope was that he would be able to calculate, under this restriction, a radiation spectrum that conformed with experiment, and that having done so he could then allow the artificial unit of energy to become infinitesimally small. This kind of method is often tried when a problem does not yield a sensible solution: solve instead a slightly different problem that does have a neat answer, and then let the difference vanish while hoping to retain the neat form of solution.

Planck's subterfuge worked, but not quite as he intended. He discovered that he could get precisely the right answer not by making his unit of energy infinitesimally small but by keeping it at some finite value. He could exactly reproduce the measured radiation spectrum, in other words, by imposing the requirement that each electromagnetic oscillation could carry energy only in multiples of some small unit; with the size of that small unit correctly chosen (and designated by a number that came to be known as Planck's constant), perfect agreement with experiment could be obtained.

At the time he made this discovery, Max Planck was forty-two years old. This is an advanced age for a revolutionary in physics.

His long years of experience in thermodynamics and statistical mechanics allowed him to see that the division of energy into small units produced the desired form for the radiation spectrum, but also made it hard for him to understand what his achievement meant. He knew, after all, that classical physics contained no such restriction, and he spent many years after his vexing discovery trying to think of ways in which this restriction might arise from some unsuspected classical argument. In this he never succeeded, and it was again Albert Einstein, ensconced in the Bern patent office, who in 1905 made the true conceptual leap away from classical physics.

Where Planck and a few others tried to derive the rule that energy came in small units as a novel consequence of classical physics, Einstein, as he did with the constancy of the speed of light, took the result at face value. Rather than searching for some mysterious reason for radiant energy to behave as if it came in small units, Einstein took the division of energy into "quanta" as a fact, and explored the consequences. By thinking of the quanta literally as "atoms" of electromagnetic radiation, he could use all the standard methods of statistical mechanics to calculate the thermodynamic properties of radiation—energy content, pressure, and so on—directly from Planck's hypothesis. This was the same sort of argument that Ludwig Boltzmann and others had originally used to support their contention that atoms of gas were not convenient mathematical fictions but real objects; they had shown that if the atoms were taken to be real objects, then all the physical characteristics of a volume of hot gas could be explained. If Einstein could do the same for electromagnetic radiation, starting from quanta, should he not then conclude that the quantization of energy was as real as the existence of atoms of gas?

Physicists did not suddenly decide en masse that Planck's energy quanta were real. What happened, as had happened with atomic theory and Maxwell's electromagnetic field, was that quantization showed itself capable of solving a great many puzzles, and once physicists began to think routinely of electromagnetic radiation "as if" it consisted of individual units of energy, it was only a matter of time before they forgot to say "as if." What is real in physics is what works: quantum theory, taken at face value, could solve problems at which classical theory balked.

Even so, quantization at first was applied exclusively to electromagnetic radiation, and it was by no means obvious in the early days of the new theory that its principles underlay all of physics. The quantum revolution was a slow one, spreading by degrees into one area of science after another. From electromagnetism the next step was into atomic theory. Once again, the impetus came from a set of experimental data that classical physics could not begin to address. At the beginning of the twentieth century, it had been known for some decades that atoms of any pure chemical element absorbed and emitted light only at specific, isolated frequencies. These frequencies had been measured and tabulated, especially for hydrogen, and ingenious physicists had found ways of arranging the dozens of measured frequencies into a small number of series, each governed by some simple numerical formula. The activity of puzzling over the published numbers and coming up with mathematical formulas based on no physical theory was derided by some as mere numerology, devoid of significance. But in 1885 Johann Balmer found a straightforward formula that contained all the known frequencies of the spectral lines of hydrogen: each frequency could be written as one fundamental frequency multiplied by the difference between the inverse squares of two whole numbers. Given the formula's simplicity in accounting for so varied a set of numbers, it was hard not to believe that Balmer had struck upon something of importance. But as with the radiation spectrum, theoretical physicists had not the slightest idea where Balmer's formula came from.

The main hindrance to theoretical understanding was the sad fact that no one had any idea what atoms were, but in the early 1900s experiments by Ernest Rutherford and his group at the University of Manchester began to establish an empirical picture of atomic structure. Rutherford took a source of alpha particles—products of radioactive decay which he knew to be tiny, heavy, and electrically charged—and arranged for a beam of them to strike a thin gold foil. Most of the alpha particles sailed through the gold foil as if through an open window, but a small proportion bounced back toward the source. In Rutherford's words, this was as surprising as if "a fifteen-inch shell fired at a sheet of tissue paper came back and hit you."[2] Up until that time, most scientists had tended to picture the atoms in a solid as if they were grape-fruit stacked in a box, filling most of space and leaving only small gaps between them. If this were so, the alpha particles in Rutherford's experiments should have shot through the atoms in the gold foil like tiny bullets through soft fruit. This was why Rutherford's discovery of occasional reversals in the paths of the alphas was so odd: to bounce back like that, the alpha particle must have run into something more solid and massive than itself.

The explanation was not hard to find. The light, negatively charged particles known as electrons had been recently discovered, and were known to be a component of atoms. Rutherford concluded that atoms must contain a dense, hard core—the atomic nucleus—about which the electrons orbited like the planets around the Sun. Most alpha particles would pass straight through such an atom undeflected, but occasionally one would hit the nucleus dead-on and be sent back whence it came. This explained Rutherford's results nicely, but it made no theoretical sense. An atom constructed like the Solar System would survive, according to classical notions, about a millionth of a second; the rapidly moving electrons would, according to Maxwell's theory, generate electromagnetic radiation, and would quickly give up their energy and spiral down into the nucleus. There was nothing in classical physics to keep the electrons in place.

The man who tackled this dilemma was Niels Bohr, a ponderous slow-talking Dane, who became the first great innovator and interpreter of what would soon be called quantum mechanics. He began by supposing that the Solar System model of the atom was essentially correct, in that electrons could be thought of as planets orbiting a heavy nucleus, but he added the condition that the electron orbits were "quantized"—that is, the electrons could circle around the nucleus not at any old distance but only at certain fixed radii. He then proposed that the emission and absorption of energy by the atom was due to electrons jumping from one orbit to another. If a unit of electromagnetic radiation of just the right frequency struck the atom, it would knock an electron from one orbit up to the next; this is why atoms absorb radiation only at certain energies. Similarly, if an electron moved down the ladder from one orbit to the next, it would send out a unit of electro-magnetic radiation of a specific frequency; this is why atoms emit radiation only at certain energies.

Here was the connection with the quantum formulation of electromagnetic radiation that Planck had stumbled on and Einstein had made rigorous. The electromagnetic energy that an atom could emit or absorb came, according to the quantum principle, in fixed units; therefore, the position of the electron orbits in the atom must correspond to those same fixed units of energy. Having grasped this basic idea, Bohr was easily able to reproduce the Balmer formula for the spectral frequencies of the hydrogen atom: a hydrogen atom has only a single electron orbiting its nucleus, but that electron can occupy one of an infinite number of orbits, each defined by a certain radius and energy. As the electron hops from one orbit to another, the hydrogen atom emits or absorbs light whose frequency corresponds to the energy difference between the orbits. Using Planck's constant multiplied by a whole number to calculate the successive orbits, Bohr was able to reproduce Balmer's formula for the set of all the frequencies of hydrogen light. By this procedure, Bohr showed Planck's hypothesis to be the basis for understanding not only the nature of radiation but atomic structure as well.

This explanation, published by Bohr in 1913, greatly impressed Rutherford, but what the Bohr model really amounted to was hard to say. It "solved" the problem of the stability of the atom only by fiat; the electrons would not lose energy and spiral into the nucleus, as classical physics required, because Bohr didn't allow them to. And it was a curiously hybrid theory: to calculate the energy of the orbits, Bohr used pure classical methods, but how an electron hopped from one orbit to another—the first example of a quantum jump—was left wholly unexplained.

Bohr's model had the undeniable virtue, however, of explaining Balmer's formula for the frequencies of light from hydrogen, and it extended the notion of energy quantization from electromagnetic radiation to the mechanical energy of electrons going around a nucleus. But the model raised as many problems as it solved. At the bottom of it all was the fact that Bohr's quantization rule for the electron orbits seemed arbitrary. In Planck's formulation, each quantum of electromagnetic radiation had an energy equal to its frequency multiplied by Planck's constant. This was a pleasingly simple idea, and therefore a credible addition to the list of fundamental definitions in physics. It seemed reasonable to expect that if Bohr's atom were to be understood further some similarly elegant formulation of the quantization rule would have to be found.

A clue to what this might be came in 1923, from Prince Louis de Broglie, a minor French aristocrat and until then a minor physicist. De Broglie wondered whether he could find, for the electrons in atomic orbit, an analog for Planck's simple proportionality between energy and frequency. The energy of each orbit was known, but since the electron was not a wave but a little tiny planet barreling around the atomic nucleus, there was no obvious meaning to be attached to an electron "frequency." Undaunted, de Broglie figured out what the frequency and wavelength would have to be, and in so doing came across a remarkable result: for the first (lowest-energy) Bohr orbit, the wavelength was exactly equal to the circumference of the orbit; for the second orbit, two whole wavelengths fitted into the circumference; for the third, three, and so on. De Broglie had, in a sense, merely played around with Bohr's quantization rule, and according to George Gamow, the physicist and raconteur, de Broglie's version of the quantization rule, obtained as it was from an argument of dubious physical significance, was pooh-poohed by more sophisticated types as "la Comédie Française."[3]

But on closer examination, there was more to de Broglie's idea than simple charm. If one thought of the electron as some sort of wave, then the quantization rule that followed made sense. A wave running in a circle around the atomic nucleus will overlap itself on each orbit, and if, in repeated orbits, the peak of the wave arrives where the trough was a few circuits previously, the net effect will be to cancel out, very rapidly, the wave motion. But if, as de Broglie supposed, the orbit is a whole number of wavelengths around, the peaks coincide on every circuit and the wave motion is reinforced. The quantization rule is now no more than the physical requirement that wave motion around the nucleus should reinforce rather than erase itself.

Neither de Broglie nor anyone else knew what this electron wave was; it was certainly not an electromagnetic wave, in the manner of Maxwell's equations, because it had to do with electrons, not light. But de Broglie's wave was soon shown to have real physical meaning. In 1927, the American physicists C. J. Davisson and L. H. Germer conducted careful experiments in which a beam of electrons was bounced off a crystal. Each atom in the crystal, when struck by the hypothetical electron wave, should set up a secondary re-radiation of electron waves; if the electron beam strikes the crystal at such an angle that the wave arrives at the second row of atoms exactly one wavelength after it has hit the first row, then all the atoms will re-radiate electron waves in phase, and the crystal as a whole will strongly reflect the incident electron wave, because all the emissions from the atoms combine. But if the angle of incidence is changed so that the wave strikes a second row of atoms half a wavelength after the first, the atoms will re-radiate out of phase; the waves will cancel each other out, and the crystal will absorb the incident electron wave rather than reflecting it. Waves bouncing off a crystal should therefore exhibit a pattern of reflection strength that increases and decreases as the angle of incidence is varied.

This indeed is what Davisson and Germer found: electrons could behave as if they were waves, not particles. At about the same time, de Broglie's hypothesis was set on surer theoretical ground with the publication by the Austrian physicist Erwin Schrödinger of the equation that bears his name. It is a wave equation; like Maxwell's equations, or the equations that describe the vibrations of a drumskin, it gives a mathematical basis for calculating undulations in a field according to forces that are applied to it. The drum analogy is useful in this case. When a drumskin is struck, it vibrates in a complex way, but mathematical analysis of the relevant wave equation shows that any vibration, no matter how complex, can be broken down into a mixture of simpler solutions, which are called the fundamental modes. The simplest mode is one in which the center of the drum rises and falls;

slightly more complicated is a mode in which one side of the drumskin falls while the other rises. Then there are modes in which the rising and falling sections are arranged in quadrants, or in concentric rings. The reason a drum produces one kind of sound when it is struck at the center and another when it is struck near the edge is that the impact of the drumstick excites different modes of vibration.

Schrödinger's equation for the wavelike motion of atomic electrons was something like this, except that instead of the up-and-down motion of a drumskin it determined the magnitude of a mysterious wave that had something to do with the electron's location. The fundamental modes of Schrödinger's equation had precisely the energies of the Bohr orbits, and the nature of the wave motions also bore some relation to de Broglie's picture, in that the most fundamental mode was a single oscillation, the second mode two side-by-side oscillations, and so on. But de Broglie's waves and Schrödinger's were clearly not the same thing, because de Broglie's waves were supposed to travel around the old-style circular Bohr orbits, whereas Schrödinger's waves represented an oscillation centered on the atomic nucleus and spreading into the surrounding space.

In the end, de Broglie's picturesque electron waves proved to be no more than a useful stepping-stone to the more sophisticated but more abstract theory proposed by Schrödinger and elaborated by many others during the late 1920s. It emerged ultimately that any electron could be wholly represented by a Schrödinger wave, which became known as the wavefunction. The rule for translating the wavefunction into classical terms was to take the square of its magnitude at any point and to think of the number so obtained as the probability of finding an electron there. From a distance, the wave would be seen as a strong, localized peak, which one could reasonably think of as a particle at a certain position, but as one got closer the wave would seem to be spread out over some small region of space, and it would make sense to think of the electron only as having a certain chance of being at a certain place at a certain time.

Thus was born, to the consternation of every generation of physics students, the notorious wave-particle duality of quantum mechanics. Electrons, represented by wavefunctions, seemed to be part wave, part particle; similarly, Einstein's use of Planck's formula for electromagnetic radiation had shown how each little piece of an electromagnetic oscillation could be thought of as an independent unit carrying a certain amount of energy—something, in fact, that could reasonably be called a particle. The "particles" of light, which had until then been regarded as the archetypal wave motion in physics, were named photons, and now electrons turned out to be waves.

Ironically, Schrödinger's innovative description of the electron in terms of an underlying wavefunction was quickly taken up by the physics community in part because it seemed to offer the promise of explaining, in old-fashioned classical terms, where quantization came from. In the case of a vibrating drumskin, we might want to call the fundamental modes, out of which all possible vibrations can be composed, the "quanta" of drum oscillations. There is nothing mysterious about these drum quanta; they come about because the circumference of the drumskin is fixed, and only a finite set of oscillations can fit within the boundary. Similarly, it seemed plausible for a time that Schrödinger's electron waves produced a quantization of the motion of atomic electrons in an entirely classical way. There is no fixed boundary within which the electron wave has to fit, but the restrictive condition is similar: if a wavefunction is to be physically meaningful, it must fall to zero at a large distance from the nucleus. Out of all the possible wavefunctions allowed by Schrödinger's equation for an atom, only some behave as required; there are solutions that become infinitely large at great distance from the nucleus, and therefore do not correspond to electrons confined to the volume of space around the nucleus. When these misbehaving solutions are discarded, the ones that remain correspond to the permitted electron orbits. It seemed possible that Planck's idea of reducing quantization to classical physics plus some extra conditions would be realized.

But this reappearance of classical thinking was an illusion. Schrödinger's waves represented, as we have seen, not the location of an electron but the probability of its location; even though the wavefunction for an electron in an atomic orbit was wholly determined by Schrödinger's equation, there was an irreducible, and decidedly non-classical, element of probability in going from the wavefunction to a prediction of what an individual electron would do in specific circumstances.

What this implied was shown in a concrete example by Werner Heisenberg in 1927. As Einstein had analyzed with infinite care exactly what was meant when two different people wanted to agree on the time, Heisenberg aimed to understand what it really meant to measure the speed and position of an electron. He imagined that the electron-measuring apparatus was a machine that bounced a photon off an electron, and recorded the direction it came off in. As noted, the photon is a little bundle of electromagnetic-wave energy and the electron is a wavefunction. When we say that the photon bounces off the electron, we mean that the two wave motions interact, but because each wave is spread out, there is necessarily some looseness in what is meant exactly by the place of collision. In other words, we cannot say precisely where the electron and photon collide, and since our measurement strategy is to discover where the electron is by bouncing a photon off it, we find that our measurement of the electron's position is inescapably compromised. To improve the accuracy of the measurement, we might try using a photon of higher energy, which has a smaller wavelength and is therefore more "concentrated" in space. Indeed, we can then say with greater precision where the electron-photon collision took place, but the problem, as Heisenberg recognized, is that by using a photon of higher energy we hit the electron harder, and therefore disturb its course. Although we have gained more precise knowledge of where the collision occurred, we have done so at the expense of knowing where the electron went next. Conversely, if we use a low-energy, large-wavelength photon so as to disturb the electron as little as possible, we can hope to find out only approximately where the electron is.

This is the notorious uncertainty principle of Heisenberg, which says in effect that the more accurately you try to measure something, the more you interfere with what you are measuring. It might be thought that the uncertainty can be eliminated by making a theoretical allowance for the collision, but this overlooks the probabilistic nature of the collision itself. A meeting between an electron and a photon is not like the classical collision between two billiard balls, which can be predicted with essentially perfect accuracy if you know exactly their speeds and paths. Both the electron and the photon are described by waves that occupy some volume of space, and when the two waves interact it is possible to calculate only the probability of one outcome or another. Heisenberg's uncertainty principle does not limit the scientist's ability to obtain accurate knowledge of the electron's position; by using photons of very high energy, the position can be deduced as precisely as one likes. But the higher the photon energy used, the more uncertain becomes the outcome of the collision. After being hit by a high-energy photon, the electron will shoot off in any direction and at any speed with almost equal probability, so knowledge of its position has been obtained at the expense of knowing its velocity.

The basic import of the uncertainty principle is that if you want knowledge of one thing, you have to pay for it with ignorance of something else. There can be, therefore, no perfect knowledge of the kind thought possible in classical physics. There must always be some sort of compromise, and it is up to the experimenter to decide which compromise to select. One reaction to this is to carry on supposing that there is a real, objective electron somewhere out there, with specific position and speed, and that the uncertainty principle prevents us from knowing exactly what its speed and position are. This is an effort to maintain the classical, commonsense idea of an objective world by imagining that the uncertainty principle puts a screen between it and us. But on reflection any such philosophy is meaningless. The only knowledge of an electron's speed and position is that which we acquire by performing some measurement, and the uncertainty principle tells us that the knowledge we acquire will be different for different experiments. It becomes utterly meaningless to insist on the existence of some absolute reality if we then have to concede that different experimenters can never agree on what that absolute reality is.

This was the true revolution of quantum theory—not the idea

that light came in packages, that energy came in quanta; not the odd fact that photons and electrons sometimes behaved as waves, sometimes as particles; but the principle that physics, at this subatomic level, possessed an irreducible element of uncertainty. The old classical electron no longer existed; instead, there was a prescription for calculating, in a given set of circumstances, what the package of properties that constitute an electron would do. Sometimes this calculation would yield an answer very much like the answer given by the classical view of an electron as a particle, but on other occasions, as with the scattering of a beam of electrons from a crystal, the answer would be more reminiscent of what classical physicists would call a wave. To argue whether the electron is really a wave or a particle is futile: an electron is an electron, and it behaves as an electron behaves. The observer may sometimes perceive wave properties in that behavior, and sometimes particle properties, and that, as Keats said, is all ye know, and all ye need to know. If we ask for more, what we are really asking for is a direct, metaphysical understanding of the nature of the electron, as if we could grasp its "true" nature with our minds and obtain some understanding of it beyond the scope of physical experimentation.

This is the second great perplexity of modern physics, the second point where common sense, learned from the everyday world, proves inadequate to the task. Just as the special theory of relativity forced us to replace the commonsense notion of fixed absolute time with a personal time we each carry around with us, so quantum mechanics demands that we discard the equally commonsense idea that an electron or any other subatomic particle is a kind of miniature billiard ball. Mechanical models of the subatomic world must be discarded along with mechanical models of the ether.

This decidedly non-classical way of interpreting quantum mechanics—ask only for what quantum mechanics can tell you, and do not expect to learn more—was evolved chiefly by Niels Bohr and his disciples, and became known as the Copenhagen interpretation. In pure form, it is a baldly rational philosophy. It declares that quantum mechanics will tell you the answer to any question you ask of a physical system, but that if you try to sneak in another question the answer to the first will change. Quantum mechanics is a set of rules for making sense of any measurements you care to make of a physical system, but absolutely forbids you from pretending to deduce from those measurements what is really "inside" the system.

Even Werner Heisenberg, one of the sternest advocates of the Copenhagen interpretation, could not altogether suppress the urge to know what is "underneath" the external world of quantum mechanics. He suggested at one time that quantum systems possessed a sort of propensity, a potential, to act in certain ways, and that depending on how a system was surrounded by detectors that propensity would be drawn out in different ways. This view was, in Heisenberg's estimation, "a quantitative version of the old idea of 'potentia' in Aristotelian philosophy. It introduced something standing in the middle between the idea of an event and the actual event, a strange kind of physical reality just in the middle between possibility and reality."[4]

It is vague yearnings of this sort, coming from such hardheaded physicists as Heisenberg, that has led to the sometime belief that in quantum mechanics physicists are beginning to stumble across mystical truths. Oriental religions may express ideas on the nature of the universe in ways which, to the Western ear, never become clear, and physicists trying to explain the workings of quantum mechanics also resort frequently to groping metaphors and analogies that don't quite succeed. But that is no reason to assume any deep connection between the two. In classical physics (in which we may now include relativity since it, too, rests on a deterministic view of the world), there exists an absolute underlying reality, amenable to measurement by ingenious scientists, which we can hope steadily to apprehend. But quantum mechanics denies the existence of any absolute reality, denies that there is a mechanical world of particles and forces existing independently of us. The world is what we measure it to be, and no more than what we measure it to be. Heisenberg, with his revived Aristotelian "potentia," was trying to explain the quantum world in terms of the classical, trying to invent an underlying reality when no such thing exists. It is impossible to explain why there is no absolute quantum reality, just as it is impossible to explain why there is no absolute Newtonian time; we had assumed these things to exist, but when our assumptions were put to the test they failed.

The practicing physicist must learn to deal with the hard rules of the quantum world, along with the disappearance of absolute space and time. In both cases, there are new rules to navigate by, and familiarity with relativity and quantum mechanics is a sufficient guide. But a loss there has surely been. The classical physicist could rely on common sense and an intuitive mechanical view of nature, because those were the foundations on which classical physics was built. The twentieth-century physicist, on the other hand, has learned to distrust common sense and intuition, and to look instead to mathematical and logical consistency for a guide into the dark and distant places—the subatomic world and the universe at the speed of light—where the frontiers of physics now press. To go further we must deliberately close our eyes and trust to intellect alone.

3

The Last Contradiction

ALBERT EINSTEIN, after formulating his special theory of relativity, turned his attention to a larger problem. The fundamental rule of special relativity is that no physical influence can travel faster than the speed of light, and one place in particular where that rule was broken was in Newton's theory of gravity, the famous inverse-square law of attraction between masses. Newtonian gravity appeared to work instantaneously. It stated that between the Sun and the Earth there existed a certain force, and that was that; if the Sun suddenly disappeared, its gravitational force would disappear just as suddenly, even though light from the Sun would still strike our eyes for another eight minutes or so, that being how long light takes to get from the Sun to the Earth. Light takes time to cross the gap, but the Newtonian force of gravity could appear or disappear instantly across any distance. The ingenious physicist could then devise a means of instantaneous communication, contrary to the basic rule of relativity, by setting in one place a mass that could be moved backward and forward and placing at some distance a balance sensitive enough to feel the gravitational attraction of the mass; according to Newtonian theory, the balance would respond at the very instant the distant mass was moved, and a signal would have been transmitted in no time at all.

This instantaneity of the influence of gravitational forces was called "action at a distance." Newton himself did not find it plausible, but his inverse-square law worked, and while he professed himself baffled as to the origin of the force that pulled massive bodies together, he declared that to worry over the cause of gravity was to go beyond the limits of what a scientist can do. It was enough that he had found the mathematical rule that described the force, and could thereby explain the motion of the planets in their orbits as easily as he explained the falling of an apple from a tree; as for anything else, he said, "Hypotheses non fingo"—I make no hypotheses. But when Einstein enshrined the speed of light as an absolute upper limit to the propagation of any physical influence, the action at a distance of Newtonian gravity became unconscionable. Finding the solution to this profound conceptual difficulty occupied most of Einstein's time and thought from 1911 until 1915, when he published his general theory of relativity.

It is a well-kept secret in the physics world that the general theory of relativity is easier to understand than the special theory, which Einstein had devised a decade earlier. The special theory fixes on the absolute and invariable nature of the speed of light, and in the process does away with Newton's absolute time, but it deals only with uniform motion in straight lines. The general theory has a much broader compass: it can cope with forces and accelerations, and in particular it addresses the question, so mysterious even to Newton, of where gravity comes from.

Einstein, more than two hundred years after Newton, explained gravity in a surprising and wholly unexpected way. Space is not flat, he decided, but curved, and bodies moving through curved space do not travel in perfectly straight lines but, like marbles rolling across an undulating floor in an old house, are steered this way and that. The ingenious part is that the curvature of space is generated by the presence of mass, so that a heavy object causes a depression into which other bodies are drawn. There is no longer a force of gravity, in the old sense. Instead, mass distorts space, and bodies traveling through curved space move naturally along curving trajectories. If we insist that space is "flat," as in Newtonian dynamics and special relativity, we are bound to attribute the curving paths to the action of some unseen force that diverts bodies from the straight lines they would otherwise follow, and this invisible force we call gravity; but if we allow space to be curved, curved trajectories become understandable without any need for a mysterious force.

And action at a distance is disposed of, too. Mass curves space only locally, where the mass sits, just as a person sitting on a mattress depresses it only where there is bodily contact. But Einsteinian space, like the mattress, has a certain intrinsic springiness, so that the local distortion is the center of a more widespread curvature. If the person moves, the curvature of the mattress changes at a speed dictated by its natural springiness; the mattress takes a little time to spring back flat when the weight is removed. Similarly, the movement of a mass in curved spacetime creates undulations that ripple off into the distance at a speed determined by the natural springiness of space. This speed is the speed of light, and if the Sun were abruptly removed from the Solar System we would, according to Einstein, feel the absence of its gravitational force at the same time we noticed the lack of light and heat.

This all sounds very satisfactory, but there is a barrier to immediate understanding. We have to learn what we mean by curved space. The idea does not come naturally to most people, and the mathematics needed to describe curvature is somewhat demanding. But once grasped, general relativity has a fundamental beauty. Best of all, it attributes structure and physical characteristics to space, so that one can think of space as a medium with properties and attributes rather than an empty, featureless void. General relativity restores to us a sense of the landscape of space, and thereby enables us to find our way around in it.

In both Newtonian physics and special relativity, space and time form an unending flat plain, featureless as a sheet of paper. In the old Newtonian view, there was such a thing as absolute time, the same for all inhabitants of this plain, and its existence meant, in effect, that it was possible to draw ruled lines across space representing twelve o'clock, one o'clock, two o'clock, and so on. Travelers in Newtonian spacetime had only to look down at their feet, so to speak, to find out what time it was, and someone who noted the time as one o'clock could be sure that it was one o'clock for all other travelers passing over the same ruled line. But in the special theory of relativity these markings are taken away, and the only time that travelers can know is the time they measure by the clocks they are carrying with them. To find out what time it is anywhere else and for anyone else, they have to communicate somehow, perhaps by means of light signals. There is nothing fundamenally difficult about any of this, but it makes life harder and is literally disorienting. We have to guard against making casual assumptions about times and lengths, and trust only what we can directly measure.

The transition from the Newtonian to the relativistic world is justifiably upsetting. It takes away our bearings and forces us to abandon a secure sense of our position in space; until we learn to be comfortable with the new set of rules, there is bound to be unease. Nevertheless, we can see why Einstein's picture is superior to Newton's. Empty space is by definition featureless, yet in the Newtonian universe there was a covert assumption that time was marked uniformly upon it, as if time were an intrinsic property of space rather than a set of measurements we impose on it. Einstein makes us face the fact that time is what we measure it to be, no more and no less, and although he gives us the means, in his special theory, to make our own maps of spacetime as we travel through it, we inevitably feel some sense of loss when the absolute cosmic map is taken away.

But in the general theory of relativity, the world is given back some independent character. The featureless world of special relativity, the unmarked sheet of paper, is replaced with a curved, twisted, crumpled surface, with valleys and mountains. This topography owes itself not to some arbitrary markings, as with Newton's sheet of ruled paper, but to the presence of physical features—specifically, to the presence of mass. Space is curved, made unflat, by mass. Mass endows space with detectable properties, replacing the feaureless void of special relativity with a landscape that has a real geography. If everyone who ever puzzled over special relativity went on to master the general theory, much of the frustrated sense that something isn't quite right with special relativity would be dispelled.

The connection between gravity and the curvature of space also brings with it a profound new understanding of what "mass" means. In Newtonian mechanics, there are actually two kinds of mass. In his laws of motion, Newton defined something called inertial mass, a quantity that represents resistance of a body to an applied force. If you want to accelerate or decelerate a moving object, or change its direction, you have to apply a force, and the greater the mass of the object you are trying to move, the more force is needed. Inertial mass is a measure of the inertia of a body: it tells you the amount of force that must be applied to produce a desired change in speed or direction.

Then there is another kind of mass, which has to do with gravity. A thrown object falls back toward the ground, pulled by the Earth's gravity. Objects of different mass fall to the Earth at precisely the same speed—a fact demonstrated by Galileo, who according to folklore dropped cannonballs of different weights off the Leaning Tower of Pisa. (Aristotle had taught that heavier objects fall faster, but, as was the rule in ancient Greece, did not bother to test this proposition, it being so obviously correct.) Newton's laws of motion stipulate that a heavier object needs a greater force to accelerate it to the same speed as a lighter one, so Galileo's results required that the force of the Earth's gravity must be exactly twice as great for an object of twice the mass. Gravity is proportional to mass, which is what Newton incorporated into his inverse-square law.

This is so well known as to be hardly worth remarking, and yet it is quite surprising; it did not have to be this way. Mass, as defined in Newton's laws, is the inertial mass, measured by the resistance of a body to an applied force, while gravity acts in some mysterious way on bodies so as to produce an attraction between them. Conceivably, Galileo might have found that heavier objects fell faster, as Aristotle supposed, or that a copper cannonball fell faster than an iron one of the same weight; we can define for any substance a gravitational "charge" that measures the intensity of its attraction to the Earth, just as we define an electric charge that measures the intensity of electrical attractions. In principle, it might have turned out that copper had a bigger gravitational charge than iron and would therefore get a stronger gravitational tug from the Earth and fall faster.

But it doesn't. The gravitational charge for copper and iron is the same, and indeed turns out to be nothing but the inertial mass. Gravitational charge and inertial mass are precisely the same thing, and we have grown accustomed to thinking of a single physical property called mass which measures both resistance to force and gravitational "attractiveness." Properly speaking, the first is called the inertial mass and the second the gravitational mass, and it is an experimental fact that the two are the same. Newton was aware that this odd coincidence stood in need of explanation, but he could provide none, and observed only that this was the way his laws of motion and of gravity worked. It was mysterious, and once again he could only declare "Hypotheses non fingo"—I make no hypotheses.

In the Newtonian way of thinking about gravity, a space shuttle orbiting the Earth is like a ball whirled around your head on a length of string. The ball would like to move in a straight line, and to keep it moving in a circle you have to apply a force to it, by means of the string. Likewise, a shuttle orbiting the Earth is prevented from flying off into space by the force of gravity. Cut the string and the ball would shoot off; suspend the force of gravity and the shuttle would sail away.

But if there is a force of gravity holding the shuttle in orbit, why do shuttle astronauts feel weightless? Compare them with a fairground astronaut in a carnival, who is being swung around in a little cabin on the end of a metal arm. The terrestrial astronaut certainly does not feel weightless, and would go flying off into the distance were it not for the confining force exerted by the cabin walls. Astronauts in space, on the other hand, do not need the walls at all: take away the space shuttle, and they would carry on orbiting just as well; if an astronaut was holding a screwdriver and then let it go, both astronaut and screwdriver would continue to orbit in tandem. If you are a fairground astronaut, you realize, if you think about it, that the pressure of the cabin wall on your body is transmitted, molecule to molecule, through your skeleton and muscles; this transmitted pressure keeps you in place, and you feel it as an internal stress within your body. But for the real astronaut, there is no such stress: arms, legs, head, and body all orbit nicely in harmony. If the links between each individual molecule in an astronaut's body could be cut, they would all continue to orbit together, each one a tiny satellite. Real astronauts stay in one piece because gravity makes each part of their bodies orbit in the same way, at the same speed. This is why astronauts feel weightless, and it happens only because gravitational mass and inertial mass are the same, so that a proportionate force retains every atom in every molecule of the astronauts' bodies in the same Earth orbit.

Once again, Einstein's genius impelled him to take things at face value. If the orbiting astronaut (Einstein actually imagined a man in a falling elevator, but the principle is the same) feels no force, it is only someone outside the shuttle, observing or imagining the astronaut's position from afar, who would come up with the idea of a force emanating from the Earth and holding in orbit the shuttle and all it contained.

Einstein's leap of imagination was this: he supposed that the astronaut's point of view, and not the observer's, was correct. Why postulate the existence of a force called gravity, when no force was felt? To say that there is no gravitational force seems at first absurd, so entrenched in our minds is Newton's way of thinking. An astronaut orbiting the Earth is clearly not moving in a straight line, therefore a force is needed. But again, Einstein would respond, what is there to tell the astronaut, who feels no deflecting force, that the trajectory of the shuttle is not in fact straight? The astronaut has no way of telling the difference between being in a shuttle orbiting the Earth or in one sailing through empty space in what an outside observer would call a straight line. From the astronaut's point of view, a circular orbit around the Earth is indistinguishable from a straight line, and Einstein's resolution of this puzzle was radical: he redefined what a straight line means. Empty space, devoid of mass, is flat space, and in it a straight line is a straight line, but around the Earth, because of the presence of the Earth's mass, space is curved, and a straight line is a circle.

This takes a little explanation. Curved space in three dimensions is not easy to imagine. But think of the shuttle orbiting above the Earth's equator, and draw the orbit as a circle on a flat sheet of paper. This is the Newtonian concept: the orbit is a circle in flat space. But now imagine, with Einstein, that the flat sheet is distorted into a bowl and that the space shuttle is running around inside of it like a ball running around inside a roulette wheel. In a roulette wheel, friction slows the ball down and causes it to fall toward the center, but the imaginary bowl of curved space is frictionless, and a ball sent running around it at the right speed will run around forever without stopping. Thus an orbiting space shuttle, were it not for the friction of the Earth's upper atmosphere acting on its surface, would remain in orbit at the same height and velocity forever.

This is an accurate but limited way of thinking of the nature of curved space, because we took a single two-dimensional slice of spacetime and, in our minds, made it curved. To construct a fully three-dimensional curved space we have to take an infinite array of two-dimensional curved slices and bend them all similarly. The only consolation to one's natural reaction that this is impossible to imagine is that it is just as impossible for physicists to imagine; they possess no special capacity of mind allowing them to envisage in a full way what space "looks like" in the vicinity of the Earth. As always, it is simply that they get used to it. Familiarity with the mathematics of general relativity, coupled with some notion that an understanding of curvature in two dimensions works in the same way in three (or four, or twenty-six) dimensions, is what enables physicists to get a grip on the Einsteinian universe in all its glory and complexity.

Einstein's theory is mathematically much more complex than Newton's inverse-square law of gravity, but once the leap of imagination has been made Einstein's gravity is conceptually much more pleasing. Instead of a mysterious force that seems to act between objects without the agency of any detectable medium, mass causes curvature, and curvature influences the paths masses take. The full measure of Einstein's revolutionary thought comes with the realization that the traditional idea of a gravitational force comes about only because we find it easier to think of space as flat, when in truth it is curved. Matter curves space, and through that space bodies take the line of least resistance. No force is needed to maintain a shuttle in an orbit around the Earth, because that circular orbit, in that curved space, is the shuttle's natural state of motion. Instead, it takes force to move the shuttle away from its circular orbit, which is why rockets need to be fired to set it on a path back down to the Earth.

Newton asserted that bodies will travel in straight lines unless some outside force moves them off it. This law can be retained if we replace "straight lines" with "lines of least resistance," which in curved space need not be straight in the geometrical, Newtonian sense. This brings us to the equivalence of inertial and gravitational mass. Why does a ball dropped from a tower fall to the ground? Because space around the Earth is curved like the inside of a bowl, and the ball rolls down the slope toward the center. But if we catch the ball and hold it aloft we need to apply a force, to use our muscles. Before Einstein, we thought that this force was needed to counteract the attractive force of the Earth's gravity, but we now understand that the force is needed because, in holding the ball stationary above the ground, we are diverting it from its preferred state of motion, preventing it from following the curvature of space toward the center of the Earth. And Newton's law still tells us that to divert a body from its "line of least resistance" a force must be applied whose magnitude depends on the mass of the body. But by mass here we mean precisely the inertial mass—the ponderousness of the body—because we are dealing with the laws of motion alone. So the force we have to apply to hold an object aloft depends on its inertial mass. Thus, to return to the Newtonian picture, the reason that gravitational mass, which determines gravitational force, is equivalent to inertial mass is because gravitational mass *is* inertial mass.

The profound realization that the equivalence of inertial and gravitational mass is not a mysterious coincidence but a necessary consequence of the dynamics of curved space persuaded most physicists that Einstein was on to something. Even so, it was important for the credibility of general relativity that some experimental evidence in its favor be found, and this was provided, in 1919, by Sir Arthur Eddington, who traveled with a small team to Principe Island, off the Atlantic coast of Africa, and came back with a series of photographs showing stars that were momentarily visible at the Sun's edge during a total solar eclipse. The pictures taken at full eclipse were compared with standard night-time pictures of the same stars, and indeed, as Einstein's theory predicted, the position of the stars in Eddington's photographs were seen to be slightly displaced, because light rays from them were steered slightly off course by the space curvature near the Sun.

The true accuracy of Eddington's test has been debated by historians, but Eddington, at any rate, was convinced that Einstein's prediction had been borne out. His success—and Einstein's—made the front pages of the newspapers; general relativity and curved space were introduced to the public, and Eddington became one of the most active proselytizers for the new theory.

Since then, other tests of general relativity have been devised and performed; nevertheless, because directly measurable effects within the confines of the Solar System are so tiny, general relativity remains a theory whose wide acceptance owes at least as much to its overwhelming conceptual power as to any empirical support. The theory, fully grasped, strikes most students of physics as something that has to be right because it seems so right. It embodies a double unification of ideas: the equivalence of inertial and gravitational mass becomes a deduction instead of an odd coincidence, and action at a distance is replaced by a gravitational influence that spreads through space at the same speed as electromagnetic radiation.

There was a further parallel between electromagnetism and Einstein's gravity. Just as Maxwell's equations yield solutions that describe electromagnetic waves traveling in space, so in general relativity there appeared to be gravitational waves, in the form of ripples in space itself, traveling at the speed of light. But while it took only a few years for the reality of radio waves to be demonstrated by Heinrich Hertz's experiment, the debate over the reality of gravitational waves has occupied decades. Even now, though there is good evidence that gravitational waves are real, there is no direct proof.

Curiously, one of those who most loudly pooh-poohed gravitational waves was the same Arthur Eddington whose eclipse expedition had persuaded the scientific world, and the world at large, that Einstein was right. Eddington declared that the waves were fictitious, a phenomenon that traveled "at the speed of thought." The reason for his doubt was subtle but real. In general relativity as in the special theory, space does not come supplied with a grid of coordinates, like a hiker's map of the mountains. Travelers must do their own navigating, and they are free to describe the geometry of some curved volume of space in whatever way seems convenient to them. For example, one would normally calculate the curvature of space around the Sun from the point of view of some imaginary person located at the Sun's center, and from this vantage point the curvature would look the same in all directions. But someone traveling through the Solar System at a considerable speed would, because of the distorting effects of motion on time and distance, see the curved space around the Sun as squashed instead of perfectly spherical.

What this means is that a given geometry of space has no unique mathematical description. More confusingly, the reverse is true: two very different-looking pieces of mathematics may turn out to describe, from different points of view, the same piece of space. General relativity abounds with confusions as a result, and Eddington's objection to gravitational waves was simply that no one had proved them to be real physical objects rather than some fairly plain piece of space seen from the point of view of an unusual observer. To be specific, one might imagine asking what a plain flat piece of space would look like to someone who was traveling through it while alternately accelerating and decelerating: would this observer construe flat space to be full of ripples?

Thus arose a long controversy as to whether gravitational waves were real or simply mathematical ghosts. A hopeful but not wholly quixotic search for gravitational waves was begun in 1961 by Joseph Weber, at the University of Maryland; he suspended an aluminum cylinder weighing a few tons in the hope that it would be set vibrating by passing gravitational waves; astrophysicists had assured him that supernovae and other violent stellar events would lead to an intermittent trickle of gravitational waves passing by the Earth. Despite all his efforts to insulate the detector from seismic disturbances and the rumble of passing trucks, what came out of his experiment was mostly just measurements of noise, and he (along with several other groups of physicists trying this experiment and variations on it) has never been able to produce unequivocal evidence of a gravitational wave.

Most physicists are now fairly confident that gravitational waves are real, but the evidence they trust is indirect. In 1974, Joseph Taylor, of Princeton University, and his graduate student Russell Hulse discovered a pulsar that was sending out radio signals every sixty milliseconds. This in itself was not remarkable. Since the discovery of pulsars in 1967 by radioastronomers at Cambridge University, a couple of hundred had been identified, and astrophysicists had convincingly explained them as rotating, magnetized neutron stars. Neutron stars are the collapsed super-dense remnants of dead stars, typically containing the mass of the Sun squeezed into a sphere about ten miles across. A neutron star that is spinning, and that has a sufficiently strong magnetic field, is seen as a pulsar, generating radio signals with metronomic regularity at intervals ranging from about one second down to a few thousandths of a second. Pulsars are the most stable flywheels imaginable; once detected and cataloged, a pulsar can be picked up again months or years later, the arrival time of its pulses still marching to the same uninterrupted beat.

Like the ordinary stars from which they are descended, pulsars are found both in isolation and in orbit around other stars. A pulsar that is half of a binary system gives itself away because its signals arrive at Earth with a regular advance and delay caused by its motion around the other star. The Hulse-Taylor pulsar was found to be in a binary system with an orbital period of only eight hours; therefore, the radius of the orbit, assuming the pulsar's companion to be something like the Sun in mass, would be about equal to the radius of the Sun. Clearly, this binary consisted of two compact objects, much smaller than normal stars, and the only reasonable possibility was that the unseen companion to the pulsar had to be another neutron star, although a non-pulsing one.

Hulse and Taylor had discovered a textbook exercise in general relativity. A neutron star is a massive, dense, featureless sphere, inducing a powerful and exactly calculable curvature of space in its vicinity. And one of these neutron stars, the pulsar itself, was in effect a clock with part-per-billion accuracy. The Hulse-Taylor binary pulsar was an ideal experimental set-up to test the predictions of general relativity in all sorts of previously untested ways; when two neutron stars are orbiting each other at close range, the differences between Newtonian gravity and general relativity become quite overt. Such a binary stirs up spacetime the way a ship's propeller stirs up the ocean, sending out gravitational waves copiously. The emission of gravitational waves drains energy from the system, and the only place that that energy can come from is the orbital motion of the neutron stars themselves: the orbit must shrink.

The gradual decay of the orbit, due to energy loss by the emission of gravitational waves, was detected by Hulse and Taylor in measurements of the pulsar's period over several years. Even though gravitational radiation from the pulsar was not directly detected, its observed effect on the binary system agreed accurately with theoretical prediction. This was enough to persuade most physicists that the strange distortions of spacetime, so fiercely debated by the physicists of this century, so contrary to the common sense bequeathed by the physicists of the last century, were real physical phenomena. General relativity had grown from an untested, arguable theory into a standard tool for astrophysicists, learned in graduate school and routinely used for solving theoretical problems and fitting theories to observations. General relativity became the new common sense. But, as the long saga of gravitational waves shows, phenomena in general relativity can be controversial, hard to formulate with the clarity theoretical physicists like, and harder still to demonstrate by experiment or observations. Of all the great theoretical constructions of physics, general relativity is the least demonstrated. It still compels one's attention for its powerful beauty of rationalization as much as for any quantitative proof of its powers.

General relativity has come to be seen not as the victor over classical physics so much as the perfection of it. Einstein unified the separate Newtonian notions of inertial and gravitational mass, and did away with the unappealing idea of "action at a distance." Causality ruled with absolute strictness: not only was there an invariable correspondence between what was caused and what caused it, but these causes could propagate their influence only at the speed of light. Instantaneity was positively forbidden. To get from nineteenth-century mechanics to the mechanics of Einstein, one had to alter one's deep-seated ideas of the nature of space and time, but once that adjustment was made the philosophical system of physics, the insistence that events in the physical world were an infinitely minute chain of consequences, entirely explicable and predictable if one only had enough information, remained as secure as ever.

But what general relativity gave, quantum mechanics took away. If general relativity perfected the classical idea of causality, quantum mechanics wholly undermined it. Probability and uncertainty were introduced into the foundations of the physical world. Probability, to be sure, was a large part of the classical theory of statistical mechanics, but it was there only as a cover for ignorance: in classical physics, every atom in a gas had a certain speed and position at any one time, and statistics was used only as a means to make predictions when not all that information was known. But Heisenberg's uncertainty principle took away the possibility of acquiring all that information—denied, indeed, that the information even existed.

The inability of the quantum physicist to know simultaneously where any particle is and how fast it is moving is pervasive and unavoidable. Radioactive decay provides a straightforward example. As nuclear structure was uncovered by Ernest Rutherford and the other pioneers of nuclear physics, it became clear that atomic nuclei are made of protons and neutrons, which are jiggling around, and that radioactivity is what happens when that jiggling becomes so great that a nuclear fragment breaks off, leaving a different set of neutrons and protons behind. Thus a nucleus of uranium shoots off an alpha particle—two protons and two neutrons in a separate little package—and becomes instead a nucleus of thorium.

An experimental fact discovered in the early days of radioactivity was that the rate at which any radioactive substance decays increases in direct proportion to its bulk: the more atoms you have, the more radioactive decay you see. This was easily understood, once the nature of radioactivity as a transformation of one nucleus into another had been grasped. If all atoms, in some given period of time, have exactly the same chance of emitting an alpha particle, then the rate at which some quantity of radioactive material decays will be in strict proportion to the number of atoms it contains. If one in every million atoms decays

every hour, then out of a billion atoms a thousand will decay in
an hour.

This seemingly innocent statement is, by the standards of classical physics, inexplicable and revolutionary. It means that there is
no way of predicting when any individual atom will decay; the
best you can hope to say is that the chance of any particular
atom's decaying is one in a million every hour. If you have an
atom that has been sitting around for a week, the chance of its
decaying in the next hour is still precisely one in a million, just as
if the atom were newly minted. A nucleus, in other words, does
not age, or grow more unstable as time passes.

This was inexplicable by classical standards, because it meant
that radioactive nuclei behaved in a manner that depended not at
all on their previous history. There was no way in which a classical physicist could measure and label a quantity of radioactive
atoms so as to predict which of its atoms would decay at which
time. In bulk, a radioactive substance behaved in a predictable
way because it contained so many atoms, but for individual atoms
the laws of radioactive decay were wholly probabilistic. This
would have disconcerted the Marquis de Laplace; he would have
wanted to know the disposition of the protons and neutrons
inside a nucleus at a certain time, and from that would have
hoped to calculate just when, as they moved around, two protons
and two neutrons would come together to create an alpha particle
that possessed enough energy to break off from the parent
nucleus. But the experimental facts of radioactive decay prove
directly that the process is unpredictable, and thus disprove the
strict determinist vision that Laplace enunciated.

Albert Einstein had been the first to show that physicists must
accept Planck's quanta of energy as real physical entities rather
than mathematical constructions, but in the end his intuition
rebelled against the probabilistic nature of quantum mechanics.
He had made causality the backbone of his conception of space,
time, and gravity, and he could not surrender to the idea that
quantum mechanics inserted into fundamental physics an
inescapable element of chance. Having established the physical
reality of quanta, he took no further part in the development of

quantum mechanics, and came finally to doubt the whole enterprise. He could hardly say that the theory was wrong, since it explained so perfectly atomic structure and the behavior of electrons, but he maintained that quantum mechanics must be incomplete, an approximate and unfinished assembly of ideas that would one day be subsumed into a larger framework. By some such means he hoped to understand the probabilistic elements of quantum mechanics as he did the statistical part of statistical mechanics—a mechanics that represented our ignorance of all the relevant facts but did not imply that the facts were intrinsically unknowable. Einstein wanted to believe that probability entered into quantum mechanics because we didn't know everything there was to know about the subatomic world.

In his later decades, Einstein inveighed against quantum mechanics and spent fruitless years trying to remove probability from it. He did not succeed, but his efforts left behind an example of how quantum mechanics seems to contradict not just common sense but his sacred principle of causality. This is the famous Einstein–Podolsky–Rosen paradox of 1935, so familiar to physics students that it is known simply as EPR. In its original form, the EPR paradox was hard to imagine and would have been even harder to test experimentally, but in 1951 the Princeton physicist David Bohm came up with a much more accessible version.[1] Bohm's formulation makes use of the quantum-mechanical property called spin, which many elementary particles possess and which can be thought of, for our purposes, as spin in the familiar sense except that it is quantized. Spin can be represented by an arrow: turn a normal right-handed screw clockwise and it will drive forward into a piece of wood; by this right-handed convention, any rotation can be described by an arrow pointing along the axis of rotation.

The idea of electron spin was devised in 1925 to explain the curious fact that when atoms were placed in a strong magnetic field certain spectral lines split into two; what had been emission at one characteristic frequency became emission at two closely spaced frequencies instead. The basic assertion made by the inventors of spin was this: an electron placed in a magnetic field tends to orient itself in such a way that its spin arrow is parallel to the direction defined by the magnetic lines of force. At about the same time, Wolfgang Pauli came up with his famous exclusion principle, which declares that no two electrons in an atom can occupy exactly the same quantum-mechanical state. Without the exclusion principle, all eight electrons in an atom of oxygen, for example, would pile into the lowest energy orbit, but because of Pauli's dictate they must instead occupy, one by one, orbits of increasing energy.

But now spin had to be taken into account. Two electrons can occupy the same orbit, as long as their spins are pointing in opposite directions; if their spin directions were the same, they would be in the same state exactly, which Pauli forbids. So electrons fill up atomic orbits in pairs, two (with oppositely directed spin) in the lowest energy state, two in the next lowest, and so on. Normally the electrons in each pair have exactly the same energy, but if a magnetic field is applied to the atom, one of the pair ends up with its spin parallel to the magnetic field, so the other is forced to point in the opposite direction. The unlucky second electron finds itself in a less favorable position with respect to the magnetic field, but it cannot switch its spin alignment, because it has to be different from the first electron. The effect of the magnetic field is to force a small difference in energy between the two electrons, and the spectral line corresponding to the orbit likewise breaks up into two.

This seems like an elaborate and somewhat arbitrary collection of facts to explain an apparently simple observation, but it was the most straightforward theory that anyone was able to think of. Before long, the reality of electron spin was experimentally established: an electron beam, presumably containing initially random spins, will split fifty-fifty into two beams when it passes through a magnetic field. Half the electrons will be "spin-up" with respect to the field, and the other half "spin-down." Moreover, if the magnetic field is set up vertically, the electron beam is split into "up" and "down" halves, but if the field is made horizontal, the beam is split into "left" and "right" halves instead. It is the action of the magnetic field that dictates which directions the electron spins must line up in.

If the application of a magnetic field to a beam of electrons is thought of as a means of measuring electron spin, then it is only the act of measurement itself that forces the electrons to choose between spin-up and spin-down—or between spin-left and spin-right, if the magnetic field is placed the other way. One might think that the electron spins in a beam of electrons are, in the absence of a magnetic field, pointing every which way, at random, but a better statement of the idea is that each electron in an unmagnetized volume of space has a wholly uncertain (in the sense of the uncertainty principle) spin direction. It is not that the spin is pointing in a certain direction of which we are ignorant, rather that the spin direction itself is undefined. Only when a magnetic field is applied is the spin forced to align in a particular way, and only at that moment is the undetermined spin transformed into a measurable quantity.

This is the kind of thinking that Einstein found repellent, and the paradox he devised with his two young Princeton associates Boris Podolsky and Nathan Rosen is addressed to precisely this point. Imagine a pair of electrons whose spins add up to zero; quantum-mechanically, spin is an absolutely conserved quantity, and if the total spin of the two electrons is zero at the outset, it must remain zero at all times. Now imagine that these two electrons are made to travel in opposite directions away from the device that created them. At some distance, a physicist uses a magnetic field to measure the spin of one of them. According to the rules of quantum mechanics, the electron-spin direction is determined only when the measurement is made, and if a vertical magnetic field is applied, the electron will be found to be either spin-up or spin-down.

So far, so good, but the paradox arises when we recall that the spin of the two electrons must add up to zero: at the moment when the physicist measures one electron, and forces it to choose between up and down, the other electron, by now a thousand miles away, is suddenly forced to be down or up—the opposite of what the first electron chose. And if things are to work out correctly, the second electron must be forced to adopt one state or the other at exactly the moment the first electron is measured. Imagine what would happen if two physicists had set up a magnetic field for each of the two electrons, and the second physicist measured the spin of the second electron a split second after the first physicist measured the first electron. Unless the result of the first measurement ("up," say) determines instantaneously the outcome of the second ("down"), it would be conceivable, given the fifty-fifty chance of an electron's spin being up or down when measured, that the two experimenters would both measure spin-up; when they got together later, they would find that the two electron spins did not add up to zero, and that would be a violation of the strict rule that quantum-mechanical spin is a conserved quantity.

In the EPR experiment, quantum mechanics seems to have acquired something analogous to the old classical idea of action at a distance, in that a measurement made at one place has had an instantaneous consequence somewhere else. It is important to realize that this is not, at least in any obvious way, the key to a means of instantaneous communication: the first physicist might tell the second physicist to take different actions according to whether the second electron is measured as spin-up or spin-down, but since the first physicist does not know, until the moment of measurement, whether the first electron is going to be up or down, there is no practical way of sending a pre-determined instruction this way.

The EPR experiment cannot be used, therefore, to transmit any information faster than the speed of light, and in that sense does not contradict the letter of the law of relativity. But to Einstein it contradicted the spirit of the law. He felt that the EPR "paradox" demonstrated the unsatisfactory nature of the foundations of quantum mechanics, he referred to this aspect of the new theory as "spooky action at a distance," and he did not want to see reintroduced into physics what he had striven so hard to banish.

As far as Einstein was concerned, the EPR experiment demonstrated a flaw in quantum mechanics. Despite Bohr, Heisenberg, and all the other members of the Copenhagen gang, who declared that nothing in quantum mechanics was real until it was measured, Einstein interpreted the EPR experiment to mean that there must be some secret way in which the electron spin was fixed, but concealed, at the outset. Einstein's preference was for what has become known as the "hidden variable" version of quantum mechanics, in which unpredictability and uncertainty represent, as in classical physics, nothing more remarkable than our ignorance of the "true" state of the electron.

This can easily seem an issue more of metaphysics than of physics. Any hidden-variable theory of quantum mechanics, if it is to work properly, must give exactly the same results that ordinary quantum mechanics provides, and to argue whether a quantity such as electron spin is genuinely indeterminate (Bohr) or determined but inaccessible (Einstein) seems pointless. A common reaction among physicists, coming across the EPR "paradox" in their studies, is to feel that there is something puzzling in it but that nothing will come of pondering it. It is generally looked on as not quite a respectable occupation for a physicist to cogitate over the unreal issues that the EPR experiment brought out. But in 1960, the Irish physicist John Bell came to the surprising conclusion that there was a practical way to tell the difference between the standard interpretation of quantum mechanics, in which the electron's spin was truly undetermined until the moment of measurement, and Einstein's hidden-variable idea, in which the spin was determined at the moment the two electrons were made, and merely revealed by the measurement. Bell showed that in slightly more complicated EPR-type experiments, the statistics of repeated measurements would come out differently in the two theories.[2] Bell's result was not particularly difficult to follow, and was surprising only because no one until then had fully thought through the implications of a hidden-variable theory. But as the career of Einstein himself illustrates, thinking carefully about the implications of a seemingly straightforward idea can be the first step into new territory.

Bell's result brought an air of physical reality into previously metaphysical discussions of the EPR experiment, but for technical reasons it was some years before anyone was able to devise an experiment that would actually test quantum mechanics in the way Bell prescribed. Finally, in 1982, Alain Aspect and his group at the University of Paris put together just such an experiment and got the result that many expected: the standard interpretation of quantum mechanics was correct and the hidden-variable idea was not. (Many theoretical physicists claimed to be unim-

pressed by Aspect's work, since it merely confirmed what everyone already knew; physicists, like the rest of us, have infallible
hindsight).

Some thirty years after Einstein's death in 1955, his fond hope
that quantum mechanics would turn out to be fundamentally
mistaken was overturned, at least for the moment. The standard
interpretation of quantum mechanics works fine, and the truly
indeterminate nature of unmeasured physical properties has to be
accepted at face value. The way to understand quantum mechanics, if it can be understood at all, is to concern oneself only with
what is measured in a specific experiment, and to ignore res-
olutely all else. Some sort of "spooky action at a distance" may or
may not be at work in Aspect's experiment, but it does no good to
worry about the matter; quantum mechanics gets the answer
right, and never mind how.

As the physicist-turned-Anglican-priest John Polkinghorne has
observed, this is a curious attitude to take.[3] The normal rule in
physics, more so than in any other kind of science, is never to
stop asking questions. The biologist, content to know what molecules do, may have no desire to understand the reasons for their
individual characters; chemists, trying to find out why some reactions between atoms occur and others do not, may be uninter-
ested in what makes atomic structure the way it is. But by tradition the physicist, having found one level of order in nature,
invariably wants to know, like the archaeologist digging down
into the remains of Troy, whether there is another, more primitive
layer underneath. Except in the case of quantum mechanics,
apparently: physicists continue to find quantum mechanics baffling (as Niels Bohr once remarked, anyone who claims that
quantum theory is clear doesn't really understand it), but they are
supposed to live with their bafflement by refusing to ask questions
that would cause trouble. It does no good, therefore, to ask what
is the nature of the spooky action at a distance in an EPR experiment; the results come out as they should, and that is enough.
John Wheeler, a student of Bohr's and a sage of modern physics,
recalls that Bell once said of himself that he "would rather be
clear and wrong than foggy and right," which surely points up a

difference in philosophy.[4] Physics has made progress because physicists have had the daring to pose new questions and seek precise answers, but as far as quantum mechanics is concerned most physicists seem content not to push their understanding too hard; they are content to live with fogginess, and only a few, like Bell, prefer to aim for clarity.

This attitude is all the more curious in that physicists know perfectly well that quantum mechanics and relativity have a basic incompatibility. General relativity relies on the old-fashioned classical notion of particles idealized as mathematical points moving along trajectories idealized as mathematical lines, but quantum mechanics forbids such thinking and insists that we think about particles and their trajectories (if we think about them at all, which perhaps we should not) as probabilities only: a particle has a certain chance of being in this region of space, and a certain chance of moving off in this direction. It is universally accepted that, in this respect, relativity is deficient and must be brought up to par with quantum theory. A quantum theory of gravity is needed precisely because general relativity uses a classical notion of objective reality that was banished long ago by quantum mechanics. But from Einstein's point of view there is also the problem that while relativity fully and completely upholds the principle of causality, in quantum mechanics there appears to be some pale remnant of the old classical bugbear of action at a distance. For this reason, it was Einstein's contention that quantum mechanics is deficient, and would have to be modified or augmented one day to include a sterner version of the law of cause and effect than it presently does. Alain Aspect's experiments prove that Einstein's invocation of some sort of hidden-variable theory will not work, but the problem that so agitated Einstein has not been entirely beaten down.

The contradictions between relativity and quantum mechanics constitute the single great question in fundamental physics today, and form the profoundest motivation for seeking a "theory of everything," but the common attitude seems to be that something must be done to general relativity to take account of the quantum uncertainty in all things. The idea of altering some part of quantum mechanics to get rid of spooky action at a distance is declared off-limits by order of the standard interpretation of quantum mechanics, which specifically forbids questions that probe beneath the surface: what you measure is all there is. It is a novelty indeed in physics that a theory should contain a stricture telling the physicist that it is no use trying to dig beneath the surface, especially when that theory, quantum mechanics, is known to be inconsistent with another fundamental theory, general relativity.

4

Botany

THE CONTROL ROOM of the proton accelerator at the Fermi National Accelerator Laboratory, situated at the edge of Chicago where suburbs are giving way to open farmland, looks like a small-scale version of mission control for a rocket launch. In a slightly darkened room, technicians, engineers, and scientists sit at a curving array of television monitors and keyboards. A few hundred yards away, and about twenty feet underground, protons and antiprotons (particles identical to protons but with negative instead of positive electric charge) are speeding in opposite directions, guided by magnetic fields and boosted every now and then by jolts of radio waves, through a stainless-steel pipe a few centimeters in diameter. The narrow pipe forms a huge circle one kilometer across—its designer liked round numbers—and therefore six and a quarter kilometers, or a little less than four miles, in circumference. The progress of the particles in the ring and the health of the electromagnets, cooling devices, and vacuum pumps that keep the accelerator working are displayed on the television screens in the control room, and from there the accelerator operators tweak the magnetic fields to keep the particles on track, insuring that they run around the evacuated ring without crashing into one another or into the walls of the tube, gaining energy on each circuit.

When the particles have been accelerated around the ring many times, and are moving at the maximum speed the machine can take them to, a small adjustment is made to the magnetic field at one point in the ring, and the protons and antiprotons smash into each other. Physicists want to know what is inside protons, for they are not, as was once thought, truly elementary particles. They are made of smaller objects, which themselves may or may not be truly elementary in the way that the ancient Greeks imagined: fundamental, indivisible, incapable of being made out of some combination of other things. Protons and antiprotons, smashed into each other with enough speed, yield up all manner of fragments—some that live only briefly before turning into something else, others familiar creatures like photons or electrons—that fly away intact from the scene of the accident. Each collision unleashes a tiny catastrophe of particles changing into other particles, with subsidiary collisions in the wreckage creating yet more particles. It is the physicist's job not only to piece together from the debris the identity of what was inside the protons and antiprotons before the collisions but also to figure out what physical laws govern the creation and disappearance of the debris.

These modern scientists, the particle physicists, have come a long way from Ernest Rutherford, who sat in the dark watching for the faint glow of alpha particles rebounding from a gold foil onto a phosphorescent screen that served as his detector. But they are Rutherford's direct descendants. Replace his alpha particles with machine-accelerated protons, replace his phosphorescent screen with a room-sized array of tiny microelectronic devices piled up column by column and row by row to form a modern particle detector, and the principle of the experiment is the same: the physicist wants to bash one particle at high speed into another, and to study the nature and trajectories of the fragments.

Rutherford's experiment, in which he bounced alpha particles off the nuclei of gold atoms, was easy to visualize: one can think of a ping-pong ball bouncing off a tennis ball. In the experiments at Fermilab and the other modern accelerators, things are not so simple. You can think of the protons and antiprotons as tennis balls, but when these tennis balls collide they explode in a shower of golf balls, marbles, and steel bearings; with a flash of light, some of the marbles subsequently shatter into dried peas. The aim of the particle physicists is to discover what happens in these collisions, and then to understand their findings. Can a golf ball create

dried peas directly, or does a marble always appear as an intermediary? If so, what law of physics demands the presence of the
marble?

Physicists do not call particles golf balls, marbles, and dried
peas, of course, but they might as well; the names they use are
neutrino, pion, kaon, hyperon, omega. Whereas in the old days
particles had just mass and electric charge, properties a scientist of
the nineteenth century would have understood, they now also
possess attributes with names like hypercharge, lepton number,
strangeness, isospin, and charm. Each particle (each weird, unfa-
miliar name) is a little package of physical qualities, uniquely
labeled by its mass, charge, charm, strangeness, and so on. Each
little package can interact with other little packages only in certain ways, according to the physical laws governing their manifold
properties.

In essence, all this is no more than a complicated and enlarged
version of the law of electrical attraction. Electric charge comes in
two types, positive and negative. Positive and negative charges
attract each other, and like charges repel. Clothes coming out of
the dryer often have some electric charge on them, and stick
together; this is a familiar, domestic kind of physics. But physicists
in this century have discovered that there are more forces
between subatomic particles than merely electrical attraction and
repulsion, and that they need more labels to represent what a
particle will do in the presence of other particles. But the bewil-
dering variety of these labels represents, at bottom, nothing more
than an attempt to describe the properties and interactions of the
multitudinous particles in as economical a way as possible.

The proliferation of particles and labels hardly seems to beto-
ken economy, but if the progress of particle physics in this century
can be represented by a single theme, it would be the search for
economy. Economy may be taken as a synonym for truth: if
physicists can arrange all their particles in a tidy way, they think
they will be closer to a true understanding of the fundamental
laws of physics than if their theories are a confection of laws with
exceptions and corollaries, and special rules for particles that
don't quite fit. Physicists persist in trying to catalog the particles as
simply as they can, with as few labels as will do the job, but they

have been constantly thwarted by the discovery of new particles, or new interactions between particles. As a definition of progress, the search for economy seems almost self-evident, but it is a modern conceit. Lord Kelvin's idea of progress, when he was trying to understand electromagnetism, was to concoct an elaborate mechanical model of wheels and axles and gears. The important thing then was not to make a simple model but to use simple ingredients and let the model be as complicated as it needed to be. The modern physicist takes the opposite view, preferring to use odd ingredients (charm, strangeness, isospin) in order to keep the model that connects them simple and elegant.

One old-fashioned mechanical notion remains, however: particle physicists indeed think of the things they are studying as particles. The casual observer can be forgiven for wondering what happened to the subtleties of quantum mechanics, so painfully developed in the early part of the century. Was not the classical notion of a particle as an object endowed with unchangeable intrinsic properties forever supplanted? Are we not supposed to think in terms neither of particles nor of waves but of quantum-mechanical constructions whose appearance is different according to the observations being made? The old classical electron, for example, was, like one of Lord Kelvin's mechanical models, a little ball with mass and charge, but the new quantum-mechanical electron is a changeable creature, defined not so much by its intrinsic properties as by the way it behaves in given circumstances. An electron is described by one of Erwin Schrödinger's quantum-mechanical wavefunctions, which says what probability the electron has of being at a certain place and having a certain effect. A collision between quantum-mechanical particles is a tangling of wavefunctions, whose outcome is another wavefunction describing how the two particles leave the scene. In a classical collision between a bullet and a bowling ball, the outcome is utterly predictable: the bullet shoots off here and the bowling ball lumbers off there. But in the quantum collision there are only probabilities; a range of outcomes is possible, some more likely than others, and what happens in any particular collision is unpredictable.

Any experiment in particle physics is necessarily an exercise in statistical physics. It is not possible to predict exactly what is going to come out of a single collision between one proton and one antiproton, even if their energies and directions were known in advance with as much accuracy as quantum mechanics permits. Equally, it is not possible to deduce from the results of that collision what exactly went into it. A single collision cannot tell the physicist very much, and it is only by compiling the results of repeated collisions, by observing how often one kind of particle rather than another is produced, by figuring out which particles tend to be produced in combination and which only when another is absent, that it becomes possible to determine in a quantitative way the physical laws of particle collisions. Quantum mechanics may predict that in a certain kind of collision one result has an 80 percent chance of happening, and another result only a 20 percent chance: if 1000 such collisions are made to happen, there ought to be 800 of the first sort and 200 of the second; if experiments yield 900 of the first sort and 100 of the second, it is time for the theoretical physicists to go back to their calculators.

The physicists at Fermilab know about quantum mechanics and wave-particle duality, of course; they learned all that in graduate school. But in the glorious mess of particle collisions, many of those subtleties are smoothed over. Werner Heisenberg imagined that when all the physics of the subatomic world was known it would be possible to depict an elementary particle in a formal way as a mathematical wavefunction containing labels for charge, mass, isospin, strangeness, and whatever other qualities were needed for a full description.[1] A complete theory of interactions between particles, including new subatomic forces as well as familiar electromagnetism, would be a souped-up version of Maxwell's theory. To find out how one particle interacted with another in this perfect theory, you would take the mathematical descriptions of the particles, plug them into the mathematical account of the interactions between them, turn the cogs, and pull out a wavefunction that described the outcome of an encounter in all its quantum-mechanical, probabilistic complexity. By making such a particle interaction occur at an accelerator, and repeating the experiment many times, physicists are able to record all the different outcomes allowed by quantum mechanics; in effect, they measure the shape of the wavefunction for that type of collision. The players in this drama are not like the old classical particles; instead they are curious mathematical formulas labeled with mass and charge and strangeness and charm. Nevertheless, each type of particle has its own fixed description, an invariable formula that can be plugged into theories to produce answers; and since you have to give this contraption a name, you might as well call it a particle.

Whether or not Heisenberg's idea turns out to be appropriate, something of this sort is what particle physicists have been pursuing since the 1930s. They want a full description of all the particles in the world and all the interactions between them, and they want it to be not a ragbag of collected rules and regulations but a systematic unified theory, a theory that possesses some aesthetics of form and some economy of ingredients which make it seem fundamental. There is no way to know in advance what this theory of everything would look like, but most physicists like to think that when they see it they will recognize it.

For most of this century the vision of completeness has seemed very distant, but on occasion it has danced before physicists' eyes. In 1932, James Chadwick, a student of Rutherford's, discovered the neutron, a particle slightly heavier than the proton and with no electric charge. Three particles—the proton, the electron, and the neutron—now accounted for the structure of atoms: electrons (described by wavefunctions) spread themselves around atomic nuclei, which are made of protons and neutrons squeezed tightly together. The positive charges of the protons are balanced by an equal number of negative charges from the electrons, so the atom as a whole is neutral; the neutron contributes mass but no electric charge.

This seemed like enough elementary particles. Even better, an equation published by the British physicist Paul Dirac in 1926 suggested that there might be only two truly fundamental particles. Dirac's equation was designed to improve on Schrödinger's equation for the behavior of electrons: it dealt with space and time in a relativistic way, whereas Schrödinger's was strictly Newtonian, and it accommodated the newly invented electron spin. In these aims Dirac succeeded, but his equation threw up a new puzzle. Solutions to it came in mirror-image pairs. For every negatively charged electron there was something behaving in much the same way but with a positive charge. What this meant was controversial. Dirac made a quixotic attempt to persuade his colleagues that the positively charged solution was nothing more mysterious than the proton—although the fact that the proton was known to be almost two thousand times heavier than the electron directly contradicted this interpretation. Dirac thought some further modification of the theory might resolve this little difficulty, and he spent some time trying to push an early unified theory that seemed to explain two of the three known particles in a single equation.

But this was a phantom glimpse of unification. Dirac's hopes were dashed in 1932, when Carl Anderson, of the California Institute of Technology, discovered a particle identical in mass to the electron but with positive instead of negative electric charge—exactly what Dirac's equation had predicted. The new particle was named the positron but also became known as the antielectron when realization dawned that Dirac's mysterious equation was telling physicists that every particle should have an antiparticle, an object of precisely the same mass but with electric charge and various other fundamental properties reversed. This straightened out the meaning of Dirac's equation but left his fleeting vision of unity in the dust, a ghost town on the plain of discovery.

In addition to the proton, neutron, and electron, there were now antiparticles for all three, and there were other theoretical puzzles that suggested that the world of elementary physics was not going to be a simple one. There was, for instance, the entirely incomprehensible fact that atomic nuclei existed in the first place. Each nucleus contains positively charged protons and neutral neutrons; between protons, squeezed together as they are in so tiny a space, there must be a ferocious electrical repulsion, which the neutrons can do nothing to counteract. Why does the nucleus not fly apart? Evidently some new force was at work. Physicists called it "the strong nuclear force," a name that reflected exactly how well it was understood: it was a force, it operated within the nucleus, and it was strong.

In 1935, the Japanese physicist Hideki Yukawa had an idea that a new kind of particle might explain the strong force. He imagined that neutrons and protons are really two species of a single kind of particle, the nucleon, and he imagined that they could interconvert by swapping back and forth a particle that came to be known as a meson (because its supposed mass was intermediate between that of a nucleon and an electron). A proton could turn into a neutron plus a positively charged meson, and a neutron could turn into a proton plus a negatively charged meson. The nucleus, Yukawa imagined, was aswarm with these mesons, so that neutrons and protons were constantly switching identities. Cocooned in a meson cloud, the neutrons and protons are forced to stay together, because each nucleon owes its existence to all the others, and none can depart the confines of the nucleus.

If Yukawa was right, it ought to be possible to extract a few mesons by crashing a particle into a nucleus with enough energy to tear away a wisp of the meson cloud. Physicists, on the trail of not just a new particle but a new force of a nature and a new kind of theory, went at it.

In the early days of particle physics, there were no big machines to accelerate particles to high energies, and physicists could not perform nuclear reactions at will. Natural sources of radioactivity, such as Rutherford and his successors used, were too tame to wreak the kind of nuclear havoc that the search for Yukawa's mesons demanded. Physicists, ever the improvisers, turned to a source of naturally produced energetic particles, the cosmic rays. These "rays," actually streams of electrons and protons with a sprinkling of alpha particles, are thought to be produced in violent astrophysical events such as supernovae, and they travel throughout the galaxy. They rain down on the Earth equally from all directions, and have huge energies, in some cases trillions of times the energy of alpha particles from radioactive minerals. Few cosmic-ray particles actually reach the Earth's surface, however. After careening for millions of years through the emptiness of interstellar space, they give up their energy almost the instant they strike the upper layers of the Earth's atmosphere, crashing into nuclei of oxygen and nitrogen and sending fragments flying.

Such a collision is the archetypal particle-physics experiment. Unfortunately, in order to make use of the particle collisions nature has laid on, the physicist must travel to a mountaintop, where the action is. Physicists did just that, lugging their cumbersome detectors to high places. By such means Carl Anderson had found the positron, and in 1936, with his graduate student Seth Neddermeyer, he found a particle more massive than the electron but less massive than the proton or neutron. This looked as if it might be Yukawa's meson, except that it was not quite heavy enough and lived too long. It lasted about two microseconds before turning into an ordinary electron, and the lifetime Yukawa wanted was a hundredth of that. In 1947, a slightly heavier and shorter-lived meson candidate was found in cosmic-ray collisions; it was dubbed the pion, and it proved to be Yukawa's meson, the mediator of the strong force, that keeps the nucleus together. But that left the first meson candidate, now called the muon, useless and unwanted but indubitably there. As I. I. Rabi remarked, "Who ordered that?"[2]

Then there was another problem, over beta radioactivity. In this form of nuclear instability, a single proton inside a nucleus changes into a neutron, releasing a positron and leaving behind a transformed nucleus with one fewer proton and one more neutron. This seemed a simple and plausible enough process, but when physicists studied it in detail, they found that the emitted positron did not carry away as much energy from the nucleus as it ought. Between the old nucleus and the new nucleus-plus-positron, a small portion of energy had gone missing.

Wolfgang Pauli suggested in 1931 that there must be another particle emitted in beta decays, one that carried off the missing fraction of energy. Pauli's idea was publicly championed by Enrico Fermi, who gave the presumed particle the joky diminutive of "neutrino," since it had to be electrically uncharged and the name "neutron" had recently been appropriated by Chadwick's discovery. The neutrino was the object of suspicion and scorn for many years. These were the early days of theoretical particle physics, and inventing a new particle with no justification except the maintenance of a scientific principle—even for the law of energy conservation—was too grandiose for many. Today, physicists invent new principles and new particles with abandon, several in a single four-page paper, but this is now and that was then.

It was more than twenty years before the elusive neutrino was pinned down. Cosmic-ray experiments had yielded up the positron, muon, and pion, along with a grab bag of other new particles, but in most cosmic-ray collisions the same few particles turned up over and over again. To search out particles that might be rarely or never produced in these collisions, new methods were needed. Cosmic rays were difficult of access and entirely uncontrollable; they could not be made to do what physicists wanted them to do.

The painstaking discovery of the neutrino made use of a new and bigger research tool: a nuclear reactor, in whose interior nuclear transformations were expected to generate neutrinos and antineutrinos copiously. The problem with neutrinos is that they were designed to be almost undetectable, because their very purpose was to steal away from the scene of a beta decay, bearing the missing energy. The only way to find neutrinos was to make lot of them and hope to catch one or two. This is what Frederick Reines and Clyde Cowan accomplished in the 1950s. Next to a nuclear reactor at the Savannah River Nuclear Plant in South Carolina, Reines and Cowan positioned their detector, a large tank of cadmium dichloride solution. The detection method, sensitive in fact to antineutrinos rather than neutrinos, was ingenious. Antineutrinos from the reactor were supposed to interact occasionally with protons present in the tank as the nuclei of hydrogen in the water molecules. Antineutrino plus proton would yield positron plus neutron, in a form of induced beta decay; the positron would then encounter an electron somewhere in the tank, and annihilate, producing a photon of characteristic energy. But positrons can be produced in other ways, so this in itself would not be a signal that an antineutrino had been found. The additional trick was that the neutron produced along with the positron would be captured by a nucleus of cadmium, creating a heavier version of cadmium that would possess some excess energy from its capture of the neutron; this energy would also be released in the form of a photon of known energy. The telltale sign that Reines and Cowan had trapped an antineutrino was to be the detection of a photon from positron annihilation, quickly followed by a second photon from the neutron capture by cadmium.

The experiment, having taken years to prepare and assemble, started running in 1952. Unequivocal detection of antineutrinos came in 1956. The Reines and Cowan experiment presaged much of what was to come in particle physics: a big machine, with a big detector embodying an elaborate identification scheme, taking years to snag a few elusive, fleeting particles. As the 1950s turned into the 1960s, experiments got still bigger, the teams of physicists and technicians larger, and results took longer to obtain. What turned modern particle physics into the vast enterprise it has now become was the invention of machines that accelerated familiar particles, usually electrons or protons, to high speeds and crashed them into targets of the experimenters' choosing. Cosmic-ray experiments had yielded over the years a great harvest, but if particle physics was to move forward it had to be brought into the laboratory, where physicists could learn how to create unusual and short-lived particles at will and force them to undergo collisions of their devising.

Modern particle physics in this practical sense begins with Ernest O. Lawrence and his cyclotron. The original cyclotron, built by Lawrence at Berkeley in 1930, was a small affair, only four inches across. It consisted of a pair of hollow semicircular metal electrodes (which became known as Ds, from their shape), placed back to back between the circular disks of a flat electromagnet. The magnetic field pervading the cyclotron forced charged particles to follow curved paths, so that they would travel around the inside of one D, cross the gap into the other D, and travel back again, repeatedly. In addition, Lawrence put an alternating electric voltage across the Ds, so that every time the charged particles found themselves in one D they would be drawn by electrical attraction to the other, but by the time they crossed, the voltage would have switched polarity, and they would be pulled back the other way. Particles were thus flipped back and forth between the Ds, gaining energy on every circuit.

The trick of the cyclotron was in giving the particles a series of little kicks, which added to their energy in increments, rather than in trying to give them a lot of energy in one go. Before Lawrence, accelerating devices were one-shot affairs, in which a thousand volts, say, of electricity would give a particle a certain amount of energy and no more. An electron accelerated from a standing start by a thousand volts acquires, appropriately, a thousand electron volts of energy; the electron volt has become the standard unit of energy in particle physics. Lawrence's four-inch cyclotron gave protons eighty thousand volts' worth of energy with an electric potential of only two thousand volts. Lawrence and his team at Berkeley, in a series of efforts that established California as a new center of American and world physics, built an 11-inch cyclotron, then a 27-inch one, and eventually a 60-inch machine.

The cyclotron was a huge success, but its reach was not unlimited. As the machines grew bigger, it became harder to build disk-shaped electromagnets of the required quality, but there was also a problem deriving from fundamental physics. As particles in a cyclotron were made to go faster, their mass increased in accord with the tenet of relativity that energy is equivalent to mass, and this increase in mass would throw off the timing between the passage of the particles from one D to the other and the frequency of the alternating voltage accelerating them. During the forties and fifties, the cyclotron evolved into the synchrocyclotron and then the modern synchrotron, and these problems were overcome.

In the synchrotron, the bulky disk-shaped electromagnets are replaced with a tube bent into a circle. Wires are wrapped around the tube, so that it becomes a string of electromagnets that can steer charged particles—protons, in the case of Fermilab—endlessly around the circle. At one point in the ring, an electric field gives the particles a little jolt of extra energy. Traveling around the ring for hours at almost the speed of light, Fermilab's protons acquire energies of not quite a trillion electron volts. The key to making the synchrotron work is that the current passing through the electromagnets must be increased as the energy of the particles is increased, which keeps the accelerated particles on the same circular track. This synchronization of particle energy and magnetic field gives the machine its name.

The replacement of the heavy disks of the cyclotron with the synchrotron's circular tube was a liberation in accelerator design. In a large synchrotron, a relatively weak magnetic field is able to control particles of huge energy. Fermilab is a kilometer across; the Superconducting Supercollider, now being built south of Dallas, will be fifteen miles across and will accelerate protons to twenty trillion electron volts—twenty times what Fermilab can manage. But the Supercollider is likely to be the last of its breed. Engineering cannot be pushed much further: a synchrotron the size of Connecticut or Switzerland seems impractical even to the most farseeing of physicists.

The advent of cyclotrons and synchrotrons made particle physics into a trawling expedition. Each new machine that was built and switched on turned up new particles at an alarming rate. Most of the new particles were short-lived things, surviving for a bare trillionth of a trillionth of a second before turning into something else, but they could be created repeatedly and predictably, and were substantial enough to be assigned masses, charges, spins, and so on. By the early 1960s, hundreds of these novel objects had been compiled. This was a great time for experimenters, and a confusing one for theorists. Enrico Fermi, who was both, is reported to have said that "if I could remember the names of all these particles, I would have been a botanist."

But untrammeled experimental discovery is not entirely a good thing; the array of newfound particles overwhelmed attempts to produce economical theories. There seemed to be many more particles than needed, and this offended the theorists' sense of parsimony; fundamental objects of nature ought to be rare and special, otherwise one would begin to suspect that they were not really fundamental at all. The urge to simplify, to economize, forced theorists into the belief that the wildly proliferating particles must all, somehow, be connected, and theoretical physics became an effort to identify patterns in the characteristics of the new particles—to find connections between them that were not obvious to the eye. Somewhere beneath this mess there had to be concealed order.

Slowly, inroads were made. The flood of particle discoveries eased, and, with time to reflect, physicists began to catch glimpses of what they were looking for. Most of the new particles, for example, were truly short-lived, lasting only a trillionth of a trillionth of a second, and it was understood that this short lifetime was due to the action of the strong nuclear force. In Yukawa's meson theory, neutrons could turn into protons, and vice versa, by pion emission; related processes would also cause rapid decay of other particles. But then there was a separate group of much longer-lived particles, meaning in this context a robust ten-billionth of a second. In 1953, Murray Gell-Mann bestowed on the latter group a new physical quality called "strangeness," and supposed that strangeness conferred exceptional longevity on its bearers. If the possession of strangeness rendered a particle immune to decay by the strong interaction, it would have to decay some other way, perhaps by means of the weak nuclear interaction, and so would live longer. For this to work, strangeness has to be something that the strong force "notices" but which the weak force does not. This might sound odd and arbitrary, but it can be translated into familiar terms: in our world there are electromagnetic interactions, which are relatively strong and depend on electric charge, and there are gravitational interactions, which are much weaker and are unaffected by electric charge. Strangeness is simply another kind of "charge" that a particle can possess, making it susceptible to a different kind of force or interaction.

This name game would have seemed like a mad and desperate enterprise except that it produced some true scientific successes. Going on from this, Gell-Mann was able to classify many of the new particles, according to mass, charge, and strangeness, into geometrical patterns of eight or ten. Some of the groups were complete, but in others there were unoccupied slots; the missing particles were sought and duly found, which persuaded physicists that the naming and labeling had to mean something.

Fermi's deprecating reference to particle physics as a kind of botany turned out to be more germane a comment than he imagined. Botanists do not mindlessly tramp through fields giving every new plant they come across a name, list each one in a catalog, and think they have done their job. They look for patterns—

for plants that resemble each other and therefore might be cousins; this is the first step in beginning to understand the evolution of plants. Likewise, it was important for physicists to begin by labeling and sorting their trove of particles according to the properties and characteristics they were able to measure; this led them to look for new particles in order to verify that the emerging patterns were more than coincidence. Only when they had verified these patterns were they in a position to begin purposeful inquiry into what the patterns meant. Like D. I. Mendeleyev, the nineteenth-century Russian chemist who arranged the increasingly numerous elements into the periodic table, particle physicists had to order their newfound world before they could understand it.

In Mendeleyev's periodic table was the glimmering of an understanding of the structure of atoms. The chemical elements can be arranged in a certain appealing and useful way because they are not a random set of objects but a set created from a limited number of ingredients, differently arranged. Likewise, the recognition that particles could be arranged usefully into patterns meant, inevitably, that there was substructure to be found. The patterns discovered in the 1950s and 1960s indicated to physicists that their particles were not truly elementary but were made up of some smaller number of more fundamental objects, combined in different ways according to a strict set of rules.

These more fundamental objects are known to us now as the quarks, and they were thought up in 1964 by Gell-Mann, at the California Institute of Technology, and, independently, by George Zweig, a postdoctoral student from Caltech working at CERN (European Center for Nuclear Research). Quarks arrange themselves in two different ways. Bound together in threes, they can make the proton (two up quarks and a down) or the neutron (two downs and an up) or a host of other particles collectively called "baryons" (from the Greek barys, meaning "heavy"). Bound up in twos, as quark-antiquark pairs (of different types, so they do not annihilate each other), they make the mesons, including Yukawa's pion, the kaon, and many more. But the tidiness of the quark model should not be taken to mean that Gell-Mann's idea was accepted the instant it was born. The chief problem was that the quarks themselves had not been seen: protons had been bashed into each other at enormously high energies, but nothing resembling quarks had ever come out of the debris. The fragments were always other baryons and mesons. "Free" quarks were not in evidence, and no one knew why. Quark theory at this time was susceptible to the same sort of criticism that was leveled at atomic theory in the early days: quarks, or atoms, might be a conveniently simple way of thinking, but unless they could be apprehended experimentally it was impossible to say whether they were fact or fiction.

There was, however, good experimental evidence that protons and neutrons had inner structure. In a series of tests done at the Stanford Linear Accelerator Center in 1967, Jerome Friedman, Henry Kendall, and Richard Taylor (all of whom belatedly won the 1990 Nobel physics prize for their labors) propelled high-energy electrons onto protons and studied how the electrons bounced off. This experiment was a direct descendant of Rutherford's scattering of alpha particles off atoms, and just as Rutherford had deduced from the pattern of scattering the presence of a tiny, hard atomic nucleus, so Friedman, Kendall, and Taylor deduced from the way electrons were scattered that protons must contain smaller objects—three, to be precise.

Still this was not taken as confirmation of the quark model. Richard Feynman, also at Caltech, analyzed the scattering results and concluded that whatever was inside the proton—he called the objects "partons"—seemed to be rattling around like marbles inside a tin can. Gell-Mann's quarks, on the other hand, were supposed to be tightly, irretrievably bound, and from a strictly empirical point of view Feynman was unwilling to say that partons and quarks were the same thing. Gell-Mann tended to be disparaging about Feynman's partons, as if his colleague were merely trying to avoid using a word he found distasteful, but there were genuine grounds for puzzlement.[3]

This experimental dilemma was eventually resolved by a new piece of theory. When the quark theory made its debut, it was largely a combinatorial idea. It showed how the baryons and mesons could be built up from a handful of quarks, but it didn't come to grips with the force that held quarks together (just as, earlier in the century, the atomic nucleus was understood to be made up of protons and neutrons, even though no one knew what held them together against their electrical repulsion). An additional complication was forced upon the quark theory by the fact that a proton contains two up quarks, along with a down, while the neutron contains two downs and an up. This is problematic, because the quarks had to obey Pauli's exclusion principle, which says that no two identical particles can exist in identical states. There had to be, therefore, some difference between the two up quarks in a proton, and a similar difference between the two down quarks in a neutron. In the end it was necessary, to explain the multitude of baryons and mesons, for each quark to come in three different varieties: this new quality came to be known as color, so that physicists now had red, green, and blue quarks on their plates.

This is beginning to seem forced, but remember as always that the names are arbitrary (the three quark types might just as well have been called "plain," "salted," and "honey-roasted"), and that once again the quantity that the quark colors denote is a generalization of more familiar properties such as mass and charge. In this case, the generalization is a continuation of what has gone before: mass is all of one type, charge is of two types (positive and negative), but quark colors are of three types. Theories can be constructed that describe the interaction between colors, just as there are theories for the interactions between charges (electromagnetism) and masses (gravity). The theory of this force, given the Latinate name of quantum chromodynamics—the quantum dynamics of the color interaction—is then supposed to explain why quarks hold together inside protons and neutrons. Quantum chromodynamics embodies the modern understanding of the strong nuclear force, and goes a step deeper than Yukawa's pion-exchange model. Pions and nucleons are themselves built of quarks, and the color interaction is supposed to explain both the construction of pions and nucleons from quarks and their interaction with each other through exchange of quarks. Yukawa's theory for the strong nuclear force thus becomes an outcome of the more fundamental theory of quantum chromodynamics.

There was still the problem of why free quarks are not seen, however, and in 1973 David Politzer, of Harvard, and David Gross and Frank Wilczek, of Princeton, came up with an innovative argument to account not only for the fact that quarks were never seen in isolation but also for the fact that the "partons" of the electron-scattering experiments seemed to be loose inside the proton. The essence of their idea is that the force between quarks of different color has a novel and perverse character: it becomes stronger as quarks get farther apart, and diminishes as they get closer together. At close range, jostling around inside a proton, quarks barely notice one another's presence and act like independent particles—which is why the electron-scattering experiments revealed loose "partons" rather than the tightly bound objects that were expected. Conversely, if one tried, with a suitable pair of tweezers, to catch hold of one of the quarks inside a proton and start tugging it out, like a dentist trying to extract a tooth, one would find the force needed to pull the quark increasing without limit. With a mighty effort, the quark could indeed be pulled away from the center of the proton, but the energy expended in tugging the recalcitrant quark out would be sufficient to create a new quark-antiquark pair between the proton and the pulled quark. To be sure, the original quark would break free—but not quite. The quark of the newly created pair would fall back into the proton, to replace the quark that had just been extracted, while the new antiquark would snap back toward the extracted quark, creating a tightly bound quark-antiquark unit: a meson, in fact.

In a similar manner, when protons and antiprotons are flung together at enormous speed, as in the accelerator at Fermilab, the energy of their collision creates new quarks and antiquarks, which catch hold of one another in twos and threes to form new mesons and baryons. Once again, no single quarks fly off.

This explanation of why free quarks were never seen was the final ingredient that made sense of the quark model. Under the rules of combination originally set out by Gell-Mann, quarks and antiquarks were the building blocks for an array of observed particles; the hundreds of baryons and mesons found at accelerators were replaced with a handful of quarks. Add to this the strong nuclear force, now explained in terms of the interaction between quarks of different colors, and it was easy to understand why quarks were never seen in isolation. But this last ingredient was a novelty in theoretical physics, and philosophically a departure from tradition. It had always been the case that demonstrating the existence of a particle meant, quite literally, catching hold of it and exhibiting it to one's fellow-physicists, as a Victorian big-game hunter would bring back a pelt from darkest Africa. But the proof of the existence of quarks was paradoxical. Politzer, Gross, and Wilczek had supplied a purely theoretical argument proving that free quarks could never be apprehended, and had thus persuaded the physics community that quarks were real. Physicists wanted to believe the quark model, because it made the world of baryons and mesons so convincingly tidy, but they hesitated over it because no one had ever seen a quark. The same sort of conundrum, a hundred years before, had made some skeptical physicists and philosophers doubt the existence of atoms: certainly atoms were a pleasing idealization, and certainly the atomic theory seemed able to explain features of the physical world which other theories could not; but if atoms themselves could not be seen, how was the truth of the theory to be judged? The virtues of atomic theory outweighed these drawbacks, as far as most physicists were concerned, and in due course atoms were directly imaged. This was the final proof, if it were needed, of the reality of atoms. But for quarks the outcome was different. The desire to believe that quarks were real was always strong, because it was such a tidy model, but in the end the quark model succeeded by the ironical trick of proving that no quark would ever be directly seen by a physicist. This liberated physicists from any need to demonstrate the existence of quarks in the traditional way.

The success of the quark model was in retrospect a turning point in the way physicists thought about the subatomic world. A complicated and disorderly "external" world, inhabited by the numerous particles that came from collisions at accelerators, could be neatly explained in terms of a smaller and much more orderly "internal" world, inhabited by quarks alone. The innovation was that it was unnecessary for the internal world to be directly accessible to the experimenter. The existence of the quarks was inferred, not demonstrated; certain suggestive experimental results, when analyzed according to the quark theory, were consistent with the fact that quarks existed but could not be captured. With the quark model as an exemplar, physicists were encouraged to think that the empirical subatomic world pictured by the operators of particle accelerators was, with all its warts and wrinkles, an imperfectly realized snapshot of a perfect underlying system. And it became the physicists' goal, in their search for a theory that encompassed all of particle physics, to find a perfect system and then to explain afterward why that perfection was hidden from us.

This search still has a way to run; physicists have by no means teased out all the concealed symmetries and patterns that govern the particle universe. There are hints and clues that an underlying system is there, but only guesses at what that system might be. The quark model filled in one corner of the whole picture, but it was hardly the end of physics. For one thing, quite a few quarks were needed. There were the plain up and down quarks, out of which the proton and neutron were made, and the strange quark, needed to account for the various long-lived particles that Gell-Mann had applied himself to. But then a fourth quark—carrying a new property called "charm"—was introduced to explain certain particles and reactions, and even after that another pair of quarks had to be invented to account for a few very rare and very massive particles. These are the "top" and "bottom" quarks, once referred to as "truth" and "beauty," to demonstrate that a few physicists knew Keats and also perhaps to suggest that these were the last word in fundamental particles. There is experimental evidence for the bottom quark but, so far, none for the top. Theories with five quarks are unappealing, however, and the search for the top quark is one of the chief goals of particle physicists today. Each type of quark comes in three colors, and each of the colored quarks has a colored antiquark to go with it, so that even within the quark model there is a mini-proliferation of particles, not as vast or disarranged as the proliferation of baryons and mesons but enough to suggest that even with quarks physicists have not yet hit ground level in the search for truly elementary particles.

And there are still a few particles that have nothing to do with the quarks. The electron, the muon, and the neutrino remain out-

side the quark model and are, as far as anyone yet knows, genuinely elementary particles, not made up of anything else. In the
1960s, a series of experiments at Brookhaven National Laboratory, on Long Island, showed that neutrinos came in two kinds;
there is the original neutrino produced in radioactive beta decays,
and another kind produced when a muon decays into an electron.
These neutrinos are not the same: the first became known as the
electron neutrino, the second as the muon neutrino. This meant
another particle, but also hinted at another pattern. The muon is
identical in all respects to the electron, except that it is about two
hundred times heavier. Now there is one kind of neutrino paired
with the electron and another kind with the muon. Is it reasonable to imagine that the muon and muon neutrino are in some
fundamental sense near-duplicates of the electron and the electron neutrino?

This tentative grouping of particles can be taken further. It
seems possible to make a habitable world out of up and down
quarks (which make protons and neutrons), electrons (to complete the construction of atoms), and electron neutrinos (to take
care of certain nuclear reactions). But, for some reason, nature
did not stop there: strange and charmed quarks make up a heavier set of particles; the muon acts like a heavy electron, and is provided with its own neutrino. Why the seeming duplication?

A third and more tentative particle family has been added,
made of the bottom quark (detected), the top quark (undetected),
a third and still heavier edition of the electron, called the tauon,
and the tau neutrino (presumed to exist, but not yet detected).
There are three families of particles, apparently all with the same
basic ingredients. Why this pattern? And why three families at all,
when the second and third manifest themselves only in exotic
processes in particle accelerators? To the eye of the physicist, this
pattern is a clear sign of some connection beneath the surface, but
no one has yet figured out what that connection might be.

The three particle families still do not include everything. The
photon, conceived by Planck and Einstein as the quantum unit of
electromagnetic radiation but now understood as the purveyor of
all electromagnetic phenomena, has been left out of this tidy

scheme, and left out with it are some new particles from the quark model. According to quantum chromodynamics, there should be eight particles called gluons (a lack of imagination is setting in among theorists) whose responsibility is to carry the strong (or color) force that binds quarks together, just as the photon carries the electromagnetic force between charged particles. Gluons are never seen openly, but, as with quarks, the evidence for their existence derives from the neatness of the theoretical models in which they participate.

One last element remains. Photons carry electromagnetism, and the gluons carry the strong force, but there is still the weak interaction, which governs radioactive beta decay and indeed any process involving neutrinos. The natural presumption is that there must be some kind of particle, a cousin of the photon and the gluon, that carries the weak interaction.

The world of elementary particles is getting to be a populous place. In the three families are the quarks (six of them, in three colors each), and the electron and its two heavier siblings, the muon and the tauon, each of which comes with a neutrino. That makes twenty-four particles, arranged into three families of eight. Outside the families are the lone photon and the eight gluons, along with one or more particles for the weak interaction—another ten particles, at least. There are patterns, to be sure, among this multiplicity, which suggests—one might even say, which proves—the existence of a substructure that particle physicists have not yet unearthed. In their efforts to dig deeper, physicists have achieved one tangible success.

During the 1970s, Sheldon Glashow, Abdus Salam, and Steven Weinberg, each contributing bits and pieces, came up with a theory in which there have to be three separate particles carrying the weak force (more complication!) but in which those three particles are intimately related to the photon (simplification!). In essence, the three weak carriers are recognized as heavy photons, and the weak force is seen as a modified form of electromagnetism. The obvious differences between the weak and the electromagnetic interactions are ascribed to the fact that the photon has no mass and the weak carriers have a lot. The mass difference between the photon and the weak carriers splits a single underlying interaction—it can now be called the electroweak interaction—into two distinct phenomena, which we observe as the electromagnetic force and the weak force.

This theory, like the quark theory, portrays apparent complication in terms of underlying simplicity. The evident lack of symmetry in the real world—the electromagnetic and weak interactions are clearly not the same—was put down to a flaw in the implementation of this would-be perfection. The role of the electroweak theory in the effort to streamline fundamental physics is in some respects debatable. More particles are needed (three for the weak interaction) and the mathematical structure of the theory itself is not entirely beyond reproach. But it ties two separate forces into a single theoretical device: the weak interaction and the electromagnetic interaction were made into one. The electroweak theory looked compelling, and it made predictions. The new particles for the weak part of the electroweak interactions had to have certain properties, and could in principle be made at accelerators. The three were called the Z^0 (which was electrically uncharged) and the W^+ and W^- (a pair of oppositely charged particles), and laboratories raced to find them. In 1983, physicists at CERN, in Geneva, found the W and Z particles. The electroweak theory was thereupon deemed correct, and the number of distinct forces in the world was officially reduced to three—electroweak, strong, and gravitational.

This is where things stand now. Electroweak unification has organized another corner of theoretical physics, as did the quark model before it, but there is still plenty left to be explained. Are the gluons related in some way to the photon and the W and Z particles? Why are there three repeated particle families, each with two quarks, an electron or its cousin, and a neutrino?

Particle physics has come a long way. Throughout this century, experimenters have found new particles and theorists have struggled to incorporate them into tidy schemes; sometimes theorists, in making tidy schemes, have predicted new particles which experimenters have then found. The present theoretical structure is both an impressive achievement, for the variety of particles and phenomena it includes, and an unholy mess, full of loose ends and replete with odd and ingenious mechanisms. Who would have thought that the force between quarks gets stronger with distance, so that free quarks can never be had? Who would have thought that a set of three "heavy photons" would make the weak interaction into a suppressed form of electromagnetism? Who would have thought nature so uneconomical as to have provided three almost identical versions of a basic set of fundamental particles?

Yet all these things have been built into the present model of particle physics because they are what is needed: they are there because they help to explain the profusion of particles and interactions revealed by particle accelerators. If particle physics is a mess, it is because that is the way the world appears to work. The quark model and the theory of electroweak unification may be neither the most obvious nor the most beautiful of intellectual constructions, but they work; they explain what goes on at accelerators, and they have made predictions that have been borne out by experiment.

And now, with particle physics in its evidently incomplete state, with many questions remaining unanswered, with patterns still unexplained, physicists are about to build (if the politicians let them) the biggest accelerators they could conceivably build. The Superconducting Supercollider is taking shape in Texas, and at CERN there are plans, not yet approved, to construct a European counterpart, the Large Hadron Collider (hadron being the collective name for baryons and mesons). Either machine may find the top quark, and should certainly pin down some missing details of the electroweak unified theory. These years of the late twentieth century may see the final triumphs of accelerator physics, but also its last hurrahs. No one expects that a significantly more powerful accelerator can be built, and no one expects that the Supercollider or the Large Hadron Collider will answer every question that remains.

Particle physicists will be obliged to continue their theoretical quest without the experimental support that has taken them this far. But they are not quite out of tricks. There is one place where energies, temperatures, and densities are high enough to probe the more subtle and rarefied aspects of particle physics. That place is the big bang—the beginning of the universe—and it is because physicists can find in the early moments of the universe what accelerators can no longer give them that they have turned, in recent years, to the study of cosmology. There is nowhere else for them to go.

II

THE
MACROWORLD

ASTRONOMERS look at the stars, and have been looking at them for thousands of years. Astrophysicists, a more recent breed, try to understand why the laws of physics make the stars shine as the do. They have known for about fifty years that nuclear reactions in the cores of stars provide the energy that makes them shine, but there is more to it than that. Stars burn their nuclear fuel at rates that vary according to their mass and composition; their temperature determines their color, blue-white stars being the hottest and red ones the coolest. Some stars in our galaxy are being born, others are about to die, and there are many more at all stages in between. Astronomers can look through their telescopes and find examples of any theoretical star—with a particular mass and composition, and at some particular point in stellar lifespan—that the astrophysicists can construct. Thus observations form a check on theory, and the scientific method flourishes.

Stars congregate in galaxies. Galaxies can be disk-shaped, with spiral arms, as ours is, or spherical and featureless. Our notion of what the universe looks like is based not on the distribution of stars but on the way galaxies are patterned across the sky. Galaxies are themselves collected into small groups and clusters, and the clusters are grouped into larger clusters. Huge filamentary chains of galaxies and clusters snake across the sky, and walls dotted with galaxy clusters separate enormous voids almost empty of galaxies. This vast construction is what we call the universe, and it falls to the cosmologists to interpret it.

Cosmologists and astrophysicists both want to use the laws of physics to construct theoretical models that explain what we see. But the task of the cosmologist is subtly different. There is only one universe before our eyes, and it is made of one particular arrangement of galaxies. We see this universe, moreover, at one particular point in its history; models of it will be very different depending on whether we think it has always looked much as it looks now, or whether it was qualitatively different in the past and will be different in the future. What we see as the universe is a single still frame from an unfolding motion picture, and the cosmologist is trying to deduce from that single frame the plot and characters of the entire movie.

This would be an impossible task altogether, were it not for a belief that the universe as a whole is governed by the same physical laws that govern our small part of it. If astronomers and astrophysicists can gather enough information on the present structure and composition of the universe, it becomes possible to use the laws of physics to wind the movie forward or backward; we can predict what the universe will look like in the near future, and we can say what it must have looked like in the near past. But cosmology depends on astronomy to provide detailed evidence of the present state of the universe, and that is not an easy business. We do not even now fully understand what is going on at the center of our own galaxy; all kinds of light and radiation is spitting from it, but the center is shrouded in gas and dust. Our knowledge of distant galaxies is much less detailed. The best picture that astronomers can assemble of how the universe looks is incomplete and patchy, deficient in important ways.

Nor are the laws of physics, applied to what we know of the universe at large, enough to build a full cosmological theory, encompassing the birth and death of galaxies as well as the history of the entire universe. The cosmologists need supplementary information, and almost always the extra ingredients take the form of philosophical preferences or aesthetic judgments. Cosmologists would like the universe to be simple, pleasing, elegant—the meaning of those words depending, of course, on which cosmologist is judging the competing theories. In the absence of any direct way of deducing how the universe evolved to its present state, a cosmologist makes some simple assumptions and figures out the consequences; if the model explains some aspect of the real universe, all is well, but if it is deficient in some way (as all theories certainly have been so far), then the original assumptions must be modified or augmented. The more complicated their models become, the less happy cosmologists are, but as astronomers have been able to chart the universe to a greater and greater depth, it has become a more complicated place. There is no guarantee that any simple model will be able to explain everything.

In cosmology, as elsewhere, modern science has pushed into areas once thought to be the exclusive domain of religion or mythology. It is an idea of the twentieth century that physics can even begin to explain the origin and fate of the universe. But if modern cosmologists tend to dismiss religion and philosophy, they have their own substitutes: they cannot deduce a complete theory of the universe from observational evidence and self-evident principles alone, so their models are always infected by their own views on what constitutes a simple, pleasing, and elegant theory. Why should the universe be any of these things? Why should it be understandable to us, for our benefit? The only real answer is no answer at all. In cosmology, as in particle physics, scientists have come to believe that a simple and elegant theory is a better explanation—closer to the truth—than a complex and arbitrary one. Simple models are credible; complicated models are more incredible the more complicated they become. Simplicity is the religion of the modern cosmologist.

But just as the particle physicists cannot see directly into the subnuclear world they wish to explore, and must therefore work indirectly and inferentially, so the cosmologists have no hope of ever seeing for themselves the creation of the universe, and are obliged to infer from what we see now what must have gone before. In cosmology, as in particle physics, experimental or observational evidence is becoming scarce, and theory is increasingly removed from what can be measured or detected; cosmology may come to be ruled more by aesthetic preference or prejudice than by the traditional principles of science. The cosmologists have achieved an impressive understanding of much of cosmic history, but have moved inevitably to the brink of that danger.

5

The Expanding Universe

E INSTEIN, like all scientists, had his share of opinions and prejudices, and it was his uncharacteristic failure to question one of these unthinking assumptions that led him to make what he came to view as the greatest blunder of his life. When, in 1917, he first used the equations of general relativity to construct a theoretical model of the cosmos, he found that no matter how he tried he could not make a static, unchanging universe; it wanted to expand. That the universe could be other than unchanging seemed absurd, and to get around this unfortunate prediction Einstein disfigured his own theory by introducing into it an arbitrary mathematical ingredient whose sole purpose was to render the cosmos static. Twelve years later, in 1929, the pioneering American astronomer Edwin Hubble published his discovery that galaxies were moving away from us at a speed proportional to their distance; the farther away they were, the faster they were moving. The universe was expanding after all, just as Einstein's equations had indicated. The conclusion that Einstein had suppressed turned out to have been a conclusion he should have stuck with; for once, his intellectual courage had failed him.

But perhaps more surprising in retrospect even than Einstein's inability to believe that the universe might be expanding was the speed with which the idea was accepted once Hubble's observations, meager as they were, were published. For thousands of years, belief in an unchanging cosmos had been automatic, reflexive. Pythagoras and some other early Greeks had believed the Earth to be round and to revolve around the Sun, but this idea had been lost, and the Greek doctrine as recorded by Aristotle, and thence fixed in the minds of subsequent generations, was that the Earth was flat and at the center of cosmos. All transient phenomena (comets and shooting stars, for example) were confined to the impermanent sublunary world, the space between us and the moon, while everything else—the ever-revolving Sun, the planets, and the wheeling dome of the stars—was beyond the moon, fixed in a state of perfect and unalterable motion. From Pythagoras and his followers the idea survived that the motion of the Sun and planets, in order to be perfect, must be circular. It was not until the time of Copernicus, who moved the Earth away from the center of the universe, and Kepler, who made the planetary orbits elliptical, that these two ideas were overthrown. Even so, the universe beyond the Sun and the planets was assumed to be fixed and permanent, unchanging, everlasting. It was not regarded as a physical phenomenon to be investigated and analyzed; it just was.

Late in the eighteenth century, the British astronomer William Herschel began to peer beyond the Solar System. An accomplished builder of telescopes, he noticed that some starry objects proved, on closer inspection, to be fuzzy patches of light—nebulae, he called them. Herschel also had the idea of surveying our astronomical neighborhood; he assumed, for the sake of argument, that all stars were intrinsically as bright as the Sun, which meant that bright stars must be close by and dim ones far away. He found that the stars in the sky were not scattered randomly at all distances from us but were confined to a roughly spherical "island universe," of which our Sun was an unexceptional member. He then leapt, imaginatively but correctly, to the conclusion that his nebulae were other island universes, so distant from our own that they could barely be seen. Few astronomers shared Herschel's belief, and debate over whether the nebulae were beyond or within our island universe was not settled even at the beginning of the twentieth century. But in 1912 Henrietta Leavitt, of the Harvard Observatory, discovered a type of variable star called a Cepheid, whose period of variation was related to its brightness; such stars could be used as measuring devices, because when the period of a Cepheid variable was measured, its intrinsic brightness could be deduced and compared to its measured brightness to yield its distance. In the early 1920s, Hubble convincingly identified Cepheids in a nebula in the constellation Andromeda and deduced it to be more than a million light-years away. Our own island universe, better known as the Milky Way galaxy, is only a hundred thousand light-years across, so the Andromeda nebula, rechristened the Andromeda galaxy in recognition of its newly demonstrated identity, had to be a distinctly separate and extragalactic creature.

The realization that stars in the universe are collected into assemblies called galaxies, which are separated by reaches of empty space, might be called the beginning of modern cosmology. It was the first convincing demonstration that anything existed in the space beyond our own immediate neighborhood, and the first indication that the universe at large was something more than a vast and featureless void. Hubble's evidence for the expansion of the universe came soon after this realization. It had been known for a decade or more that some of the nebulae exhibited "redshifts"—that is, light from hydrogen gas within them, emitted at certain well-known frequencies, was observed by astronomers on Earth at significantly lower frequencies. This was most easily attributed to a version of the Doppler effect, by which the sound of a retreating police siren falls in pitch, and seemed to mean that some of the nebulae must be moving away from us. By 1929, Hubble had detected variable stars in several of these nebulae and thereby established their distances, proving them to be galaxies beyond our own; moreover, he found that redshift increased in proportion to distance, and if redshift was taken to indicate that the galaxies were speeding away from us, then Hubble's observations led to the conclusion that the universe as a whole was in a state of general expansion. Only a decade after the Andromeda nebula had been identified as a galaxy in its own right, the notion of an expanding universe had supplanted three millennia of fixation on an eternal and unchanging cosmos.

It is a coincidence that both the observational evidence for an expanding universe and the theoretical means to comprehend it came into the world at the same time. Hubble's measurements of the recession of the galaxies were possible because of the recent

construction of the giant Hooker telescope, with its 100-inch mirror, on Mount Wilson, outside Los Angeles. But it is doubtful
whether observational evidence with such revolutionary consequences would have been so quickly taken up had not the general theory of relativity, still in its infancy, been available at about
the same time. General relativity was the first physical theory
capable of making broad statements about the universe. Newton's
inverse-square law of gravity could be used to predict the motions
of any number of bodies, provided that all their positions and
speeds were known at the outset, and indeed it would have been
possible to construct, from Hubble's measurements, an entirely
Newtonian model of an expanding universe. But this would have
been no more than a post-hoc description of facts; there is nothing in Newton's theory to say that the universe must expand, only
that it can.

Hubble himself was cautious about leaping from the measurement of galactic redshifts to the idea of an expanding universe. In
1935, he warned that "nebular redshifts are, on a very large scale,
quite new in our experience, and empirical confirmation of their
provisional interpretation as familar velocity shifts [that is, as cosmic expansion] is highly desirable."[1] But this was the caution of a
pure observer, a simple collector of facts; most theoretical physicists saw that Einstein's theory of gravity, unlike Newton's, was
able to predict expansion, rather than merely describe it, and
jumped with little hesitation to the grand interpretation of a universe that was increasing in size. Without Hubble's observations,
most scientists would no doubt have followed Einstein in deciding
that something must be wrong with a theory that predicts an
expanding universe. But it is equally likely that if Hubble's observations had come along before general relativity had been established, many scientists would have decided that Hubble must
have made a gross error to arrive at so ridiculous an idea as cosmic expansion.

The reason Einstein's theory of gravity is able to make statements about the universe as a whole is that it is a more encompassing theory than Newton's. It takes away the mysterious gravitational "force" between two bodies and replaces it with the more
subtle but more understandable idea that the presence of mass

makes space curved, and that the curvature of space makes massive bodies roll toward each other. But now the structure of space as a whole becomes an essential ingredient in the analysis of any physical problem where gravity is important—which certainly includes the problem of figuring out why galaxies should be moving away from each other.

Einstein's equations of general relativity link two complicated mathematical quantities: one is derived from the curvature of space and the other is derived from the placement of matter in that space. These two quantities are required to be equal, so the curvature of space is directly tied to the way matter is distributed within it. Matter creates curvature, which then tells matter how to move; matter, therefore, pushes matter around. This is why general relativity is conceptually simple but in practice difficult: in a system of massive bodies—galaxies, for example—the curvature of space makes each body move in one direction or another. But as the bodies move, they also change the way space is curved. Curvature makes bodies move, and moving bodies make a changing curvature, and thus the way a system changes with the passage of time results from an intricate feedback.

There is a corollary to this. In Newtonian theory, space is flat, featureless, and invariable, and matter can be placed in it as one likes, but in general relativity it is not possible to regard the geometry of space separately from the distribution of matter within it, because these two things correspond to the two sides of Einstein's equations and have to agree at all times.

Einstein's equations are difficult to solve, and the only way to make progress is to put in some simplifying assumptions. This is the first opportunity, within modern cosmology, for theoretical prejudices to make an appearance. The standard assumption, or prejudice, is that material is, on the average, scattered uniformly throughout space. Obviously, this is wrong. The material of the universe is contained in the stars and gas of individual galaxies, which are dotted sparsely across otherwise empty space. But the hope is that, overall, galaxies are placed randomly through the universe, so that an astronomer looking at the sky through a badly out-of-focus telescope would see a more or less equal blur of light everywhere. If we make, for the time being, the assump-

tion that matter is uniformly distributed, then it becomes possible
to find solutions to Einstein's equations, and to see if those solutions represent something like the universe we live in. If so, the
assumption of uniformity is evidently not so bad, and progress has
been made.

If this simple assumption is made, then the equations of general relativity admit only a few solutions. When Einstein first tried
to construct a model for the universe, he wanted one that was not
only uniform in space but unchanging in time; space, he thought,
should be filled with matter at the same density everywhere, and
that density should be the same in the infinite past and into the
infinite future. This was Einstein's updated version of the age-old
doctrine of a constant, unchanging cosmos. The problem was that
his own theory would not let him do what he wanted. A universe
constant both in time and space was not a solution of the equations of general relativity—unless the density was exactly zero. In
other words, you could have a perfect unchanging universe, but
only if there was nothing in it. As soon as some matter was introduced, the influence of that matter on the curvature of space, and
the feedback of the curvature on the distribution of matter, forced
the universe to change somehow with time, to be inconstant.

So perturbed was Einstein by his inability to construct a static
cosmological model that he decided to alter his theory to get
around the restriction. To the two terms in his equations, the one
representing curvature and the other representing matter distribution, he added a third, of no obvious provenance. This extra
term became known as the cosmological constant. It allowed Einstein to put matter into the universe, but then in effect to cancel it
out: the cosmological constant was a sort of "antidensity" that
enabled the curvature to behave as if the universe were empty
even though it actually contained matter.

There was never any physical argument in favor of the cosmological constant, only the philosophical desire to make a static
universe. The fact is, however, that if Einstein had not introduced
the cosmological constant someone else surely would have. General relativity is the simplest of a mathematical class of theories in
which matter determines spacetime curvature, and addition of
the cosmological constant is the simplest modification that can be

made to it. Theoretical physicists being what they are, many modifications to general relativity have been proposed over the years, sometimes as attempts to forge a link between gravity and some theory of particle physics, sometimes out of nothing more than mathematical play. None of these alternatives has prospered, but once published they acquire a sort of independent existence and are press-ganged into service on occasion to explain the odd cosmological observation.

Cosmologists certainly want to keep their models as simple as possible, and therefore resist the introduction of theoretical ornaments like the cosmological constant unless there seems to be a good reason for doing so. On the other hand, there is always the temptation to fiddle around; if unadorned general relativity seems to be a passably accurate explanation of the facts, why not try altering the theory a little to see if the accuracy can be improved? The compromise between the urge to keep theory simple and the urge to explain every last detail is one that can be decided only by the aesthetic judgment of the scientific profession. There is no rationally correct answer to this sort of dilemma, and the best that can be hoped for is that as both theory and observations improve in accuracy and detail a single satisfying model will begin to stand out from the pack. As long as new evidence can be brought to bear, this process works. But when the evidence is in short supply, the search for a good theory can reach an impasse, in which the ideas promulgated by different cosmologists depend on their irreconcilable personal opinions of what is right and proper.

If the cosmological constant is left out, the universe must expand. Under the standard cosmological assumption that matter is spreadly evenly throughout space, three types of universe become possible. If the density of matter is the same everywhere, then the curvature of space must also be the same everywhere, and there are only three possibilities: space can be positively curved, negatively curved, or flat. "Flat" space—space with zero curvature—is the familiar space we all grew up with, in which parallel lines never meet and the sum of the angles within a triangle is 180 degrees. A flat universe is exactly that: a flat, blank sheet of paper extending to infinity.

Positively curved space is harder, but by no means impossible, to imagine. In two dimensions rather than three, the surface of a sphere such as the Earth is a space of constant positive curvature. A big triangle made by drawing lines on the surface of the globe to connect, let us say, London, New York, and Rio de Janeiro, is not the same as the triangle made by drilling straight holes through the body of the Earth between the same points. If we were to add up the angles at the corners of the surface triangle, we would find their sum to be a little more than 180 degrees, but we customarily account for this by imagining that the surface triangle is not really a triangle but a peculiar curved three-dimensional construction. If, on the other hand, we are in the position of creatures strictly confined to the surface of a globe and unable to drill through its body or move off its surface and survey it from outside, then lines drawn on its surface connecting three points make the only geometrical object we can reasonably call a triangle. Triangles, we would find, always have angles which add up to more than 180 degrees. This would be a geometrical fact of life, and from it the mathematicians of our species would deduce that we live in a positively curved space.

A universe of positively curved space is like the surface of the Earth, but in three dimensions rather than two. Those of a mathematical bent like to think that once they have understood something in two dimensions they understand it in any number of dimensions. This may be true in an intellectual sense, but it is not the same as being able to visualize, in some familiar way, what a three-dimensional curved space really looks like. Cosmologists and physicists can't visualize a three-dimensional positively curved space any better than anyone else. They simply think of a two-dimensional space (the surface of the globe) and persuade themselves, from their understanding of the mathematical description of curvature, that they know how a three-dimensional curved space works. When they say they understand what three-dimensional curved space is, they mean that they have become familiar with it, and that its properties no longer perplex them. But they cannot draw it or otherwise demonstrate it to you, except to indicate an ordinary globe and ask you to imagine the same thing with an extra dimension. The surface of a globe has, for instance, a finite area, but no boundaries (one can walk forever without encountering an edge), and the cosmologist similarly understands that a positively curved universe has a finite volume, although a space traveler can travel forever without reaching an edge. For this reason, such a universe is called "closed"; you can travel forever, but you cannot get out of it.

Negatively curved space, the third cosmological possibility, is a different matter altogether: it cannot be visualized at all, and one has to rely completely on the mathematical description. In a negatively curved space, the angles of a triangle add up to less then 180 degrees; their sides, loosely speaking, are concave. It is possible to imagine this: think of a "saddle" shape, as of a mountain pass where a road climbs up through a valley between two mountain peaks and then descends. At the point of the pass, the altitude of the road is at its highest but it sits nevertheless at the lowest point of the valley between the two peaks. In two-dimensional terms, space here is negatively curved; a triangle drawn around the apex of the pass will have angles that add up to less than 180 degrees. A flat sheet of paper laid down cannot cover the ground without tearing, whereas a sheet of paper applied to the surface of a globe (positively curved space) will buckle up because there is more paper than you need.

The problem with this mental picture is that even after wrestling with it you have only imagined a space that is negatively curved at one point. For cosmology, what is needed is a space that is negatively curved, to the same degree, everywhere. Like the flat universe, it goes on forever and has infinite volume, so both types are often called "open" universes. But unlike flat space, no picture of a continuous negatively curved space can be rendered, either as a drawing on a sheet of paper or as some kind of sculpture. To put it in formal terms, a positively curved space can be "embedded" in flat space of one higher dimension: it can be constructed as the two-dimensional surface of a three-dimensional sphere, and thus more easily apprehended. But (as mathematicians can prove) there is no way to similarly embed a negatively curved two-dimensional space in the familiar three-dimensional world, and we, since our mental apparatus seems to be attuned to flat space, cannot visualize such a thing.

Nonetheless, cosmologists have to deal with it, and they learn to use their mathematical understanding of negatively curved space as a blind man uses a stick to feel his way about. There has often been a sort of sentimental preference among cosmologists for universes that are flat or positively curved; they can rationalize this preference, but at bottom it may be no more than an expression of the fact that they, like the rest of us, find negatively curved space difficult to navigate.

These three types of universe, positively curved, flat, or negatively curved, cannot be static, but must expand or contract. Hubble's observations indicated that galaxies were moving away rather than approaching, so our particular universe must currently be expanding. The three types of universe differ, however, in their long-term behavior. A positively curved universe is closed in time as well as in space; it expands, grows to a maximum size, then begins to contract. Flat and negatively curved universes, on the other hand, are open in time and space, and expand forever. The difference between them is that a flat universe expands less quickly as it grows, and, hypothetically speaking, coasts to a stop in the infinitely distant future, whereas a negatively curved universe is truly open, and keeps expanding with some finite speed forever. Whether the universe is closed, flat, or open rests on a comparison between two numbers: the rate of expansion, which is known as Hubble's constant and is given by the ratio of a galaxy's recessional velocity to its distance, and the average mass density of the universe, which is the number of grams of stuff per cubic meter there would be if all the mass in all the galaxies were smeared out uniformly through space. For a given density, an open universe expands faster than a flat one, and a flat one expands faster than a closed one, so accurate knowledge of these two numbers determines which sort of universe we live in. But this question has not yet been answered. Neither distances to galaxies, nor their masses, nor their distribution through space has been easy to reckon from observations, and although the range of possibilities has been narrowed in seventy years of observational effort, all three options are still allowed by the numbers. The flat universe represents the borderline case between the other two, and we know that we are near that line. But we cannot tell on which side the real universe falls.

Whether our universe is open, flat, or closed is a question that ought to be a matter for simple measurement. Astronomers should be able to look at the universe and the things in it and tell us which one we inhabit. But the measurements have proved difficult and contradictory, and no clear answer has been found. Hubble's first measurements of the expansion rate yielded (by extrapolation back to the point when the universe would have been of zero size) an estimated cosmic age of around a billion years. The age of the Earth, however, as deduced from the radioactive decay measurements to which Lord Kelvin objected so strongly, was closer to four or five billion years. The Earth could hardly be older than the universe that contained it, and this outright contradiction suggested that perhaps the cosmological models derived from general relativity were altogether wrong. It also was the motivation for a contending model of the universe, the steady-state theory, which for a couple of decades attracted a strong minority interest among cosmologists.

In the steady-state universe there was supposed to be no overall cosmic evolution. Stars would certainly form, burn up their allotment of hydrogen, and die, and galaxies would go through cycles of stellar death and rejuvenation, but in general the universe would look the same at all places, and at all times. This was what Einstein had originally tried for, but abandoned as soon as Hubble had convincingly established the cosmological expansion. Thomas Gold, Hermann Bondi, and Fred Hoyle, the founders of the steady-state universe, were in this respect arguing for a return to earlier, simpler times, when the unchanging nature of the cosmos was held to be a more or less self-evident principle, but they seized on the discrepancy between the age of the Earth and the early values of Hubble's constant as an empirical reason for preferring the steady-state conception, according to which the universe always had and always would expand at the same rate.

Consequently, there was no connection between the expansion rate and the age of any particular object in the universe, so the age discrepancy was immaterial.

There was an obvious problem: how could the universe be both expanding and unchanging? To get around this difficulty, Hoyle and his colleagues came up with a peculiar piece of physics, which they called "continous creation." To insure that the universe, even while expanding, remained the same, they proposed that new matter was constantly appearing out of the vacuum of space to fill the gaps left behind by the receding galaxies. The amount of matter involved was quite small (an atom per cubic meter per billion years), but the damage done to conventional physics by this proposition would have been enormous had it turned out to be true; continuous creation would not sit easily with the law of conservation of energy, among other things.

Admirers of steady-state cosmology tried to overcome the arbitrariness and ugliness of continuous creation by arguing that their universe was based on philosophically superior foundations. To make theoretical cosmologies of any sort, they argued, one had to have some confidence that the laws of physics that apply here on Earth apply equally throughout the universe. However reasonable an idea this might seem, it is nevertheless an assumption that cannot be directly tested until we can travel to the Andromeda galaxy and repeat the classic experiments of physics there. Hoyle and company circumvented the problem by saying that the laws of physics are guaranteed to be the same everywhere, and throughout time, if we dictate that the universe itself is entirely uniform and unchanging; in that case, our place and time in the universe is wholly indistinguishable from every other place and time, and the laws of physics have to be the same at all places and times. Bondi, Gold, and Hoyle elevated this definition into something called the Perfect Cosmological Principle, which declared that the universe, if it is to be understandable, must be constant not just at all places, but at all times. There is a certain seductive appeal in this argument, but in fact it is a modern cousin to mystical assumptions of the Pythagorean type, according to which the universe makes sense only if it conforms to "divine" dictates of order. Like the old Greek arguments, the Perfect Cosmological

Principle, considered from the pragmatic physical point of view, is fundamentally silly.

The first part of the Bondi, Gold, and Hoyle argument is correct: it is indeed an essential assumption that the laws of physics are constant throughout space and time. But where the Perfect Cosmological Principle misleads is in its suggestion that when the universe changes, the laws of physics will change too. General relativity says precisely the opposite: if the theory is assumed to act with unerring equality for all of space and time, the universe models it predicts are necessarily evolving universes. One might imagine, for example, that as the universe expands, the force of gravity gets weaker, and indeed theories that have this property have been invented. But these theories predict cosmological models different from the ones predicted by general relativity, and, so far, general relativity looks to be the better bet for explaining what we see in the universe around us.

The crucial point is that the universe, in the modern way of thinking, is no more than a physical phenomenon governed, like everything else, by the immutable laws of physics. The laws of physics are not tied to one particular kind of universe, or to one particular place in the universe, and do not care about us and our cosmic disposition. The Perfect Cosmological Principle embodies what might be called a psychologically Earth-centered universe: it admits that the Earth is not physically at the center of everything, but nevertheless tries to tie the state of the universe as a whole into our particular experience here on Earth. And in the end, Bondi, Gold, and Hoyle had to invent continuous creation in order to make the steady-state universe work; the Perfect Cosmological Principle was supposed to tie cosmology and the laws of physics into one harmonious package, but putting it into practice forced its adherents to devise, out of the blue, a bizarre piece of physics supported by no empirical evidence, either on Earth or elsewhere in the cosmos.

The steady-state universe was promulgated beginning in the late 1940s, but by the early 1950s the age problem, which partly motivated it, had been resolved. Hubble's first estimates of the cosmological expansion rate had been based on his estimates of the distances to nearby galaxies, which relied on his identification of Cepheid variables in them. It turned out, however, that Cepheids came in two types. The ones Hubble had measured were intrinsically much brighter than he had supposed, which meant that the galaxies he had measured were much farther away than he had thought. The universe suddenly became about ten times bigger, and therefore about ten times older. The age of the universe was now closer to ten billion years than to one billion, and all was well again.

This took away the single empirical prop for the steady-state universe, but the theory had by then acquired a life of its own. Although it has never caught on in a big way, the steady-state model has exercised a fascination over a handful of cosmologists despite the persistent lack of observational evidence in its favor. The yearning of Pythagoras for a constant, ordered cosmos has not disappeared. But constancy and order are not the same thing. The order that modern cosmologists desire is in the laws of physics, and while they expect the universe to follow those laws, they do not expect constancy of the cosmos, and do not feel that a changing universe is flawed, or aesthetically inferior to an unchanging one.

This is the ultimate emancipation from the doctrines of ancient philosophers and the medieval church. Copernicus broke away from old thinking by arguing that the position of the Earth in the universe was nothing special. Modern cosmology teaches that the universe itself is nothing special. It must conform to the laws of physics, but those laws exist independently of the particular form the universe takes. Physics determines how the universe came about and must evolve, not the other way around. Modern cosmology is founded on the belief that a full understanding of the laws of physics will make the universe understandable too; there should be no extra pieces of physics dreamed up to make cosmology work according to some preconceived aesthetic notion.

The mathematics and geometry of expanding universes were first formulated in 1922, by Alexander Friedmann, a Russian mathematician and meteorologist. But it was, ironically, a churchman, the Belgian Catholic priest Georges Lemaître, who first took seriously the idea that atomic physics might be able to explain the evolution of the universe. Since galaxies are now retreating from each other, there must have been a time in the past, Lemaître reasoned, when they were more or less touching; before that time they cannot have existed in the form we now see them. But galaxies consist of stars with a lot of empty space around them, so Lemaître could push the cosmic clock back farther and think about a time when the universe was so small that the stars were squashed up against each other. He could go farther still: stars consist of atoms with empty space around them, so at some still earlier moment the atoms themselves must have been pushed up against each other. In 1927, when Lemaître first began to think along these lines, the neutron had not yet been discovered, and nuclear physics had barely been explored. Lemaître had no way of knowing what the universe might have looked like before the time when atoms were squeezed up against each other; there were no laws of physics that would take him farther back than this. From the viewpoint of general relativity, the equations of an expanding universe could be extrapolated back to a single mathematical point, a singularity, when everything in the universe was compressed to infinitesimal size. But this singularity would also have to have infinite density, and this neither Lemaître nor anyone else could begin to think about. He therefore imagined that if the universe had a beginning it was as a "Primeval Atom," which, as it expanded, fragmented into stars and planets and galaxies scattered uniformly throughout space, making the universe we now inhabit.

From the 1930s well into the 1960s, cosmology was almost wholly an observational science, its main aim being to determine the Hubble constant and measure the cosmic density, thus to establish whether the universe was closed, open, or flat. Speculation as to its origins seemed like speculation indeed, hardly real physics. But Lemaître's Primeval Atom was not entirely forgotten. In the late 1940s, the Russian émigré George Gamow, a gadfly of physics who had made contributions to nuclear physics and the study of radioactivity, and Ralph Alpher and Robert Herman, of the Applied Physics Laboratory at Johns Hopkins University, tried to apply some simple notions of nuclear physics and thermodynamics to cosmology. They were able to refine Lemaître's

Primeval Atom idea, arguing that at very high cosmic density electrons and protons would be squeezed so close together that their mutual electrical repulsion would be overcome and they would combine to form neutrons.[2] From this dense neutron fluid the universe would be born, somehow evolving into what we now see. Gamow knew that an isolated neutron decays, after about ten minutes, into a proton and a electron, and imagined that, as the universe expanded, its initial complement of neutrons would be transformed into protons and electrons, which would presumably pair up to make hydrogen, the basic ingredient of the cosmos. Neutrons, protons, and electrons were thought at that time to be fundamental particles, not divisible into anything smaller; it made no sense to imagine what might have preceded the neutron fluid. This was as far back as Gamow and his colleagues could go; as it had been for Lemaître and as it still is today, extrapolations backward in time to the earliest moments of the cosmos have to cease when there is no physics with which to do the extrapolation.

Alpher and Herman added another ingredient to this picture of the very early universe. It seemed unreasonable that the universe should have started out consisting of nothing but cold neutrons; if the neutrons possessed some energy, and were crashing into each other, their motion would have created heat, and this heat would have manifested itself by the presence of photons of electromagnetic radiation. Having argued that the early universe must have been hot, Alpher and Herman were able, remarkably, to calculate just how hot it must have been; assuming only that the neutrons exchanged energy freely with the photons, they could relate the temperature of the radiation to the density of the neutron fluid. As the neutrons were beginning to decay, the temperature would have been some billions of degrees, and if the universe was hot and dense in its earliest moments, they concluded, some remnant of that primordial heat must linger, even though the universe had expanded and rarefied since its birth. In a paper unnoticed at the time, and forgotten until others rediscovered the same idea seventeen years later, Alpher and Herman predicted that the universe today should be filled with feeble radiation, its intense heat reduced to a mere five degrees above absolute zero (−273°C). This is cold indeed, and the photons of electromagnetic radiation with a temperature of five degrees would not be visible photons or even infrared ones, but radio waves of about 1 cm in wavelength, not unlike the microwaves that permeate a microwave oven.

In 1965, a group of theoretical cosmologists (there was beginning to be such a species) at Princeton independently came to the same conclusion, realizing that there should be a sea of cosmic microwave photons. Around the same time, at Bell Laboratories, not many miles away, two radioastronomers named Arno Penzias and Robert Wilson inadvertently detected the microwaves.[3] They were conducting tests on a large microwave receiver, which had been designed for satellite communications but which they hoped to use to study radio emissions from the Milky Way. They found to their consternation that no matter what they did and no matter where they pointed the receiver, there was a constant drizzle of background noise that threatened to upset the observations they wanted to make. They tried repainting the receiver and scraping out the pigeon guano, thinking there was some simple mechanical interference in the device, but the noise persisted. Its temperature, deduced from the intensity of the radio signal they were finding, was about three degrees above absolute zero. Eventually, they heard that a group at Princeton, motivated by some theoretical work in cosmology, was actually trying to build an instrument to look for what they had already found. A few phone calls, and all was settled: the universe had been been very hot, and now it was very cool; Penzias and Wilson had found the microwave background, the faint remainder of the primordial heat.

This discovery was the single step that transformed theoretical cosmology, the science of the birth of the universe, from speculation into real science. It gave credence to the seemingly farfetched idea that a straightforward, almost naïve application of laboratory-tested nuclear physics to the early moments of the universe could generate predictions that astronomers could then check. With the detection of the microwave background, the "hot big bang" devised by Alpher, Herman, and Gamow was no longer speculation but a scientific model (the name "big bang" was invented by Fred Hoyle on a BBC radio program as a derisory name for theories beginning from some sort of dense primordial state). As it turned out, Gamow's particular idea of a neutron fluid was some way off the mark, but the general idea was not. By the 1960s, the physics of the atomic nucleus was better known, and it was understood that at temperatures of several billion degrees a medium of neutrons alone would not be stable. At these temperatures, neutrons would turn into protons and electrons and neutrinos, neutrinos and protons would recombine to recreate neutrons and positrons, positrons and electrons would annihilate to create photons; all possible reactions and interactions would boil and bubble the cosmos into a seething ocean of neutrons and protons, electrons and positrons, photons and neutrinos. In this updated version of Lemaître's Primordial Atom, the universe is born from a hot broth of nuclear and other elementary particles. It was becoming genuinely possible—within the realm of reason—to think about the very birth of the universe as an event dictated by the fundamental laws of physics. If it was true that the early universe had been hot and dense, then there must have been a time when the only relevant rules were the rules of nuclear and particle physics. From this mixture of simple ingredients somehow the universe we now see was born. How to get from a boiling pot of particles to an array of galaxies scattered through the largely empty heavens was a complete mystery then, and a substantial mystery even now. But there was reason to think that a solution to that mystery could be found, within physics alone.

The conception of the early universe as a hot broth of nuclear particles led directly to the second enduring triumph of big-bang cosmology. In the late 1940s, when Gamow came up with his notion of a primordial universe made solely of neutrons, the origin of all the chemical elements in the universe—the carbon, oxygen, and nitrogen from which we are made; the silicon, iron, nickel, and aluminum that constitute the planets; the rare trace elements (gold, silver, uranium) that turn up here and there—was far from clear. Gamow had hoped that as the cosmic neutrons decayed and turned into protons and electrons they would initiate a chain of nuclear reactions that would build up the observed quantities of the elements, but this was far too ambitious a hope; the universe was cooling too quickly for any but the simplest of nuclear reactions to occur, and therefore only the lightest elements could conceivably be made.

Fred Hoyle, on the other hand, wanted the cosmic elements to be made entirely by nuclear reactions in the interiors of stars; this suited the steady-state theory. To attack this formidable problem he enlisted the help of William Fowler, an American astrophysicist with a background in nuclear physics, and two British astronomers, Geoffrey and Margaret Burbidge, both of whom were enthusiasts for the steady state.

The result of their endeavors was a single long paper, published in 1957, which by itself established the science of nuclear astrophysics. This work did almost everything Hoyle wanted. It demonstrated indeed that nuclear-fusion reactions within stars created energy by turning hydrogen into heavier elements, and that in the process carbon, oxygen, nitrogen, neon, magnesium, aluminum, and so on were produced in the sort of quantities that astronomers measured throughout the galaxy. But stars could not make enough helium. After hydrogen, helium is the second most abundant element in the universe, accounting for about 35 percent by weight of all the stars and gas in our galaxy and others. Of this 35 percent, perhaps a third could be accounted for by the burning of stellar hydrogen into helium, but the remainder seemed inexplicable. The conclusion has to be that stars must have contained something like a quarter of their weight in helium even when they formed.

It soon became apparent that the big-bang model could provide the 25 percent of helium that eluded Burbidge, Burbidge, Fowler, and Hoyle. When the universe was hot and dense and seething with particles, protons and neutrons could not combine to form the nuclei of elements heavier than hydrogen, for the simple reason that it was too hot: any neutrons and protons that happened to get together would be instantly blown apart by all the hugely energetic particles bashing into them. But as the universe expands and cools, neutrons and protons can stick together, through chance collisions, and stay together: atomic nuclei can be born, and can survive. There is a brief window of opportunity, lasting for just a few minutes in the early life of the universe, when it is neither so hot that nuclei are blown apart nor so cool that they are immune to further reaction and change. In this brief interval, neutrons and protons pair up and swap partners and form groups, and at the end of it the universe contains 75 percent hydrogen, 25 percent helium, and a trace or two of other light elements.

During the late 1960s and early 1970s, this nucleosynthesis argument was elaborated into a persuasive pillar supporting the big bang. Huge computer programs were written, combining experimental data on nuclear reaction rates with a precise model of the expanding universe when it was a couple of minutes old and ten billion degrees in temperature, to calculate exactly not just the abundance of helium but that of deuterium (about one part in a hundred thousand); helium-3, which has one neutron less than the normal form of helium (comparable to the deuterium abundance); and lithium-7, the only element heavier than helium that is produced in any significant quantity (about one part in ten billion). Over the years, an impressive agreement has arisen between the predicted element abundances emerging from the computer programs and the measured values coming from the astronomers.

The successful explanation of the light-element abundances, coming close after the discovery of the microwave background, settled the big-bang versus steady-state argument conclusively in favor of the big bang. Those few who cling to some alternative (a descendant of the steady-state universe, some form of "cold" big bang, and so on) do so in the face of all the quantitative evidence, and must contrive explanations for those things that the big bang takes care of automatically. In recent years, the success of the standard big-bang model has reached new levels of refinement. In 1982, a husband-and-wife team of French astronomers, Monique and François Spite, published for the first time an estimate of the abundance of lithium in a population of very old stars; the abundance they measured was a few parts in ten billion—exactly what the computer calculations had predicted. This was the essence of modern physical cosmology: into a universe model derived from general relativity, the rules of nuclear physics were mixed, and what emerged from the mixture was a precisely calculated set of elemental abundances, which astronomers could then check by diligent observation. The scientific method worked: theory led to predictions that led to observational tests. It was surely time to declare theoretical cosmology a real science, solid and quantitative.

And there was a more subtle triumph in the 1980s. The elementary particles, physicists had by this time decided, were arranged in three "families": the first containing the up and down quarks, the electron, and the electron neutrino; the second containing the charmed and the strange quark, the muon, and the muon neutrino; and the third containing the bottom and top quarks, the tauon, and the tau neutrino. Why there should be three such similar families the physicists could not say, but there they were. Moreover, although these three families appeared to be sufficient to explain the known elementary particles, there was no guarantee that there might not be a fourth family, or a fifth, or an infinite number, each containing particles more massive and inaccessible and exotic than the previous one. Cosmologists, it happened, were able to say something on this issue, whereas the physicists could not.

From the cosmologists' point of view, the important point about these particle families is that each of them contains a neutrino. If the early universe really were hot and dense and seething with reactions between particles, then each of these neutrino types should have been present in equal numbers. The neutrinos sit there inertly because they interact so weakly with other particles, but collectively they carry energy, and contribute thereby to the overall mass content of the universe. The more neutrino types there are, the greater the mass energy in the universe at a given temperature, and the faster, according to general relativity, the universe then expands. This has an effect on the amount of helium produced, since the window of opportunity when nuclear reactions can proceed arrives earlier in cosmic history. If there are more neutrino types, more helium is manufactured.

In the first attempts at theoretical cosmology, a knowledge of nuclear physics or particle physics was applied, to the first moments of cosmic history, a helium abundance was deduced, and comparison was made with observation to see if the model worked. In the 1980s, cosmologists were sufficiently confident to reverse the argument. A knowledge of the physics of the early universe, coupled with precise measurements of the helium abundance, could be used to fix the rate at which the universe must have expanded in its very early moments, which fixes in turn the number of neutrino types. At first, this argument was good enough only to limit the number of neutrino types to five or less, but as better data and calculations became available, cosmologists concluded, in 1983, that the number had to be less than four.[4] Since three neutrino types were known to particle physicists, the cosmological argument implied that there were three and only three types of neutrino, and therefore only three families of elementary particles. This was a turnaround: cosmologists were now using their science to tell particle physicists something the particle physicists did not know.

Because it was indeed an audacious step, the cosmological "proof" that there were only three particle families was not quite taken seriously by physicists. Particle physicists devised their own tests, which required measuring the decay rate of the W and Z particles of the unified electroweak theory. The more kinds of neutrino there are, the more ways the W and Z can decay into other particles, and the faster they therefore decay. In 1989, this measurement was done both at CERN and at the Stanford Linear Accelerator Center, and the answer agreed with what the cosmologists said: there were indeed three neutrino types, and no more. Having achieved their own independent answer to the question, particle physicists were disposed to admit that the cosmologists' convoluted argument had been right after all. Still, the cosmological reasoning was of a new sort, and it was genuinely important for physicists to satisfy themselves of its trustworthiness.

For cosmologists, the argument that there are exactly three types of neutrino was more than just another success of the big-bang model. It reinforced the point that cosmology was a real science, with its own methods and ideas, and proved that it could contribute to other areas of science rather than borrow from them. The prediction of the abundances of helium and the other light elements and the prediction that there should be a universal three-degree microwave background were made by taking established laws of physics and putting them in a cosmological context; the light-element abundances came from applying nuclear physics to cosmology, and the microwave background came from elementary thermodynamics. In both cases, the physics was standard; it was the use in cosmology that was new. But the prediction that there were just three neutrino types was different: the cosmology was in this case a standard, widely accepted piece of physical theory, and it was the particle physics that was unknown. The success of the cosmologists in this one case was proof, if proof were needed, that cosmology was genuine science, in the traditional sense: there was a model that made predictions, and ways to test those predictions by experiment or observation.

Particle physicists were happy because they had discovered a new way of testing their theories—throw them into the cosmologists' mill and see what comes out—and cosmologists were happy because they felt themselves no longer to be the junior partner in an unequal alliance. The relationship between the two disciplines was transformed from a convenient acquaintanceship into a true marriage, in which each side needed the other.

Particle physicists have begun to realize that they are running out of traditional ways to test their theories. The discovery of the W and Z particles of electroweak unification demonstrated that the practice as well as the idea of unification between fundamental forces was an attainable goal. But to go beyond this, in terms either of refining electroweak unification or moving on to a grander synthesis in elementary physics, has begun to look problematic. There is the construction of the Superconducting Supercollider, twenty times more powerful than Fermilab's accelerator, to look forward to, but a twentyfold increase in energy is only a toddler's step into the great uncharted territory of particle physics. Theories have been formulated that deal with energies and particle masses trillions of times too great for the Supercollider to see, and it is inconceivable that any terrestrial experiment could probe the landscape the theoretical physicists are beginning to nose into.

Cosmology becomes, therefore, the last resort of the particle physicists. The only place that can supply the energies, densities, and temperatures needed to explore fundamental physics in its entirety exists far back in time, near the beginning of the universe. The big bang has become the ultimate particle accelerator, and just as physicists now study the debris from particle collisions to deduce the inner structure of particles, so cosmologists now study the debris of the big bang, scattered across the skies as galaxies, to deduce what the birth of the universe must have looked like.

The difference, of course, is that experiments in particle physics can be designed, altered, and adjusted to suit the needs of the experimenter, whereas the universe is a fait accompli, beyond astronomers' control. Cosmology is a science of pure observation and deduction, and there is, moreover, only one universe we can observe. Cosmologists must make their assumptions about the state of the early universe, and particle physicists can then test their models against the cosmologists' theorizing. But if the answers do not come out right, does one tinker with the cosmology or the particle physics? The answer is that one must tinker with both, since theories neither of the earliest moments of the universe nor of the deepest levels of subnuclear structure are directly accessible to the experimenter or observer; both are tentative, and open to debate.

Cosmology and particle physics have become, in this sense, a conjoined subject, since it is no longer possible to have absolute certainty in either. Theories of cosmology and particle physics, as they are used to reconstruct the big bang, must be developed and tested in parallel, with assumptions on each side laid open to question when problems arise. There is nothing wrong with this—and, indeed, it is inevitable. There is simply no way for cosmologists and particle physicists to move forward except together. Particle accelerators can probe a certain distance into the subnuclear world, but no farther; cosmologists can see the universe only as it is now, not as it was in the beginning. Full understanding of the fundamental particles and forces or of the birth of the universe will be achieved, if it is ever achieved, only through a long and indirect chain of reasoning from what happened a long time ago, under circumstances we can barely imagine, to what is before us now.

6

Grand Ideas

As COSMOLOGISTS and particle physicists, arm in arm, began to push back to the earliest moments of cosmic history, they solved problems but also found themselves facing new questions. What had seemed in the early days of cosmology to be self-evident facts loomed as riddles demanding answers. Why, for example, does the universe contain matter but, as far as anyone has been able to tell, no antimatter? The sky is full of galaxies containing stars made of protons, neutrons, and electrons, but there are no antigalaxies containing antistars made of antiprotons, antineutrons, and positrons. This is a simple enough observation, but it seems to contradict the almost sacred principle that the laws of physics are exactly the same for matter and antimatter. If the universe came into being through the action of fundamental physical laws, as physicists and cosmologists would like to believe, there would be no way of choosing matter over antimatter: the universe ought to contain equal quantities of both. Until particle physicists began to poke into cosmology, few people had worried about this question, but once noticed it was too blatant a puzzle to be ignored. The joint ventures of cosmologists and particle physicists were evidently not all going to be instant success stories.

At the bottom of this conundrum is a single number. The amount of matter in a certain volume of space can be measured by the number of protons and neutrons—together called baryons—it contains, and the ratio of this number to the number of microwave-background photons in the same volume remains the same even as the universe expands. The number of photons per unit volume belonging to the microwave background is very accurately known, as it depends only on the measured temperature of the background, but the number of baryons per unit volume, which is directly related to the average density of matter in the universe, can be estimated only approximately. To the best accuracy that anyone can tell, there are something like a billion or ten billion photons for every baryon. The problem of how matter came to be in the universe is then twofold: why is there matter at all, and, since there is matter, why is there one baryon for every billion or so photons? Physicists, to feel that they understood cosmology, wanted answers to both questions.

Neutrons and protons, being large composite particles made of quarks, do not retain their individual identity as the cosmic clock is turned back. When the universe was about one ten-thousandth of a second old, its density was so high that protons and neutrons would have been squashed up against each other. At earlier times and higher densities than this, the baryons could not have had any sort of individual identity; there was instead a stew of quarks, interacting directly with each other, and the cosmic mix of quarks must have contained the right ingredients to make, as the universe cooled, lots of protons and neutrons but no antiprotons or antineutrons. The quarks, in other words, must have been marked by some tag—call it "baryon number"—that led, when the universe was sufficiently cool, to the emergence of baryons but not antibaryons, and this non-zero baryon number must have existed in some form all the way back to the beginning.

The pragmatic attitude is then to say that nothing made baryons appear in the universe, rather that the universe happened to be born with a positive baryon number. The presence of matter but not antimatter had then to be regarded as one of the initial conditions of the universe: it was just there from the outset. This may sound like sweeping the problem under the carpet, but in fact it was an honest confession of ignorance. The universe from one minute old to the present day could be understood in terms of familiar physics, but more knowledge was needed to go back earlier, and as long as the required new physics was not to hand, worrying about what went on back then seemed futile.

The opposite point of view was that if the symmetry between matter and antimatter was taken seriously as a fundamental principle of physics, something odd was undeniably going on. Regardless of any new and yet to be understood processes of physics, the principle of matter-antimatter symmetry dictated that a universe containing matter but no antimatter could not arise in any natural way. To suppose that the universe started off with matter, in the proportions of one unit of baryon number for every billion or so photons, seemed not only absurdly arbitrary but positively incredible, if the symmetry of the laws of physics meant anything. To some cosmologists, the only rational solution was to insist, despite all the evidence, that the universe really did contain equal amounts of matter and antimatter.

Thus arose the notion of matter-antimatter symmetric cosmology. Like the steady-state universe, it gained a small but persistent following because of its supposedly superior logical principles, but as with steady-state cosmology, physics had to be manhandled to preserve the logic. There is no immediate evidence of antimatter in the universe, so proponents of these models had to figure out where it could be concealed. Our galaxy, certainly, cannot have any antistars or antigas clouds in it, because if an antistar ran into a gas cloud, or a star into an antigas cloud, there would be a spectacular and unmissable display of high-energy gamma rays from the annihilation of protons and antiprotons, neutrons and antineutrons, and electrons and positrons. No such phenomena are in view, and enthusiasts for the symmetric cosmologies had to argue that matter and antimatter must be divided up at least on the scale of galaxies.

But that is not possible either. Our own galaxy has two irregular satellite galaxies, the Large and Small Magellanic Clouds, and among the three there is some tenuous gas held in common. If either of the Magellanic Clouds were made of antimatter, we would know about it. In larger clusters of galaxies, too, some gas is held in common, blasted out by supernovae to collect at the center of the cluster. This cluster gas, because it is hot, emits X-rays, but here again there is no sign of the characteristic gamma rays that would erupt from the encounter between gas from a

galaxy and antigas from an antigalaxy. Even the largest superclusters of galaxies must therefore be either all matter or all antimatter.

To make a plausible case for a matter-antimatter symmetric
universe, it was necessary to imagine that matter and antimatter
were separated somehow on the very largest of scales, corresponding to clusters and superclusters of galaxies. This raised new
problems. The early universe cannot have contained equal
amounts of matter and antimatter all mixed up together, because
if it did, the baryons and antibaryons would have been annihi-
lated, and only radiation would have been left. So the advocates
of symmetric cosmologies had to assert that something separated
matter from antimatter and hived them off into separate regions
of space early in cosmic history. Like proponents of steady-state
cosmology, who had to come up with the strange idea of the
"continuous creation" of matter in otherwise empty space to keep
their cosmology alive, so proponents of symmetric cosmology
were forced to appeal to some previously unknown physical
mechanism that could separate matter from antimatter.

No one was ever able to think of a satisfactory way of doing
this, and for a fundamental reason. The separation must certainly
have occurred by the time of nucleosynthesis, so that regions containing baryons could create helium. But that is so early in cosmic
history that whatever process caused the separation, restricted as
it must be to work at no more than the speed of light, would have
segregated the universe into regions of matter and antimatter
each containing no more than about one solar mass. Yet a single
galaxy contains about a trillion stars, and a rich cluster may contain thousands of galaxies. What this means is that no physical
process could possibly have separated matter and antimatter
regions on the scale even of galaxies, let alone clusters of galaxies.
Proponents of symmetric cosmologies faced a dilemma. If they
wanted to turn the idea of matter-antimatter symmetry into an
inviolable cosmological principle, they ran straight up against
another equally inviolable principle, that no physical influence
can travel faster than the speed of light. For most cosmologists,
this argument was enough to put the lid on symmetric cosmologies, and those few who persisted with them were forced to

invent fanciful physical processes, abandoning known physics in order to preserve an appealing principle. But using unknown physics to explain the universe was exactly what they had set out wanting to avoid.

The consensual view was to grant that the universe was fundamentally asymmetric, containing matter but no antimatter, and to regard the cause of the asymmetry as, for the time being, an insoluble problem. This unadventurous, wait-and-see attitude was in the end the correct one, because in the 1980s it was realized that new theories in particle physics might provide the answer to this puzzle.

By the mid 1980s, there was a general feeling that the unified electroweak theory of Weinberg, Salam, and Glashow was correct. In tying together electromagnetism and the weak nuclear force, which governs beta radioactivity, it predicted the existence of three new particles, the W and Z bosons, which were duly discovered at CERN. Unification of the forces was by this time held to be the ultimate goal of theoretical physics, and the success of electroweak unification encouraged physicists to move on to the next step: this was to tie the electroweak interaction to the strong nuclear force, which governs the interactions of the quarks and the construction of baryons from them. It was quickly recognized that theories built to the same plan as the electroweak theory, but operating at much higher energies, might accomplish this "grand unification." The practical difference was that whereas the W and Z particles of electroweak unification were accessible to particle accelerators, the corresponding particles—imaginatively called X and Y—that derived from grand unification were far too massive ever to be seen in any laboratory or machine. Grand unification belonged to an entirely new regime of energies and masses in particle physics, and the X and Y particles would have to be about a hundred trillion times heavier than the W and the Z. It was not remotely thinkable that such particles would ever be apprehended on Earth.

At this point, particle physicists turned to the cosmologists. Grand unification might not be testable on Earth, but if the early universe was what cosmologists said it was, there would have been a time, a tiny fraction of a second after the big bang itself, when the X and Y particles of grand unification were as abundant as neutrons and protons at the time of nucleosynthesis. Grand unification, therefore, might have had some effect on the evolution of the very early universe, and might conceivably have left some traces behind in the appearance of the universe today. One such trace was the presence of matter but not antimatter in the modern universe. Without grand unification, the quarks, and the gluons that held quarks together, were the sole province of the strong interaction, and particles like electrons, muons, and neutrinos took no part in it. But with grand unification, this barrier was brought down; there were new interactions (connected intimately with the X and Y particles) that could mediate between the previously separate worlds of the strong and the electroweak interactions. These hypothetical grand unified interactions could turn baryons into non-baryons, and vice versa.

One of the most surprising implications of grand unification was that it took away the absolute stability of the proton. The proton is the lightest of all the baryons, and the reason that it was thought to be incapable of decaying into any other particle was that the conservation of baryon number was held to be sacrosanct. The only way a proton could decay would be by turning into a collection of less massive particles, and since there is no less massive particle that carries baryon number, there is nothing for the proton to decay into. Therefore, the proton cannot decay. But within grand unification there are new kinds of particle transformations, some of which could turn quarks into non-quarks and vice versa. It becomes possible, if grand unification is correct, for two of the three quarks inside a proton to turn into an antiquark and (for example) a positron; the antiquark clings to the remaining quark to form one of the mesons, and the net result is that the proton has changed into a meson and a positron. In the process a single unit of baryon number vanishes altogether.

Now the cosmological puzzle looks very different. As long as it was believed that baryons could not be created or destroyed, the cosmic preponderance of baryons over antibaryons was a basic and unchangeable property of the universe. But as soon as reactions that create or destroy baryon number are allowed, the issue is suddenly not "fundamental" any more. It becomes conceivable that reactions between particles can create baryons from nothing, so that one can imagine a universe that begins with no overall baryon number (the aesthetically preferable choice) and acquires baryon number along the way.

In fact things are not quite this simple. Grand unification introduces new interactions that create and destroy baryons, but these processes have counterparts that create and destroy antibaryons. Any theoretical physicist writing down a grand unified theory would begin, for the sake of symmetry, by making all these processes and their counterparts equal. But in the cosmological context this puts us back where we started. If, in a universe containing no baryons at the outset, every process that creates a baryon is matched exactly by a process that creates an antibaryon, then we will still end up with a universe that contains equal proportions of matter and antimatter. This will not do. What is required is to add an element of asymmetry to the grand unified theory, so that processes that create baryons are a little more effective than processes that create antibaryons. Then, early in cosmic history, baryons and antibaryons are both made from nothing, but slightly more baryons are made than antibaryons. At some later time, antibaryons and baryons annihilate, but a few baryons are left over. The extent of this asymmetry is directly related to the present ratio of baryons to photons in the universe. To get one baryon for every billion photons now, there must have been, a long time ago, a billion and one baryons for every billion antibaryons.

Formally, there is no difficulty in adding this one-part-in-a-billion element of asymmetry to grand unification, so that matter comes out of the big bang slightly ahead of antimatter. Equally, however, there is no compelling reason to do it except to explain the cosmological preponderance of matter over antimatter. In terms of theoretical purity, the required asymmetry is a small disfigurement of grand unification, as the cosmological constant was a disfigurement of general relativity. The difference is that the one disfigurement appears useful, even necessary, whereas the other is not.

Grand unification seems to provide an explanation of the matter-antimatter asymmetry of the universe, but it is an ambiguous victory. The original idea, which was to assess the merits of grand unification by seeing what cosmological implications it contained, has been reversed: grand unification has been adjusted to provide an explanation for a cosmological puzzle—why there is matter but no antimatter in the universe—and the adjustment thereby made cannot be independently tested.

We begin to see the dangers of mixing particle physics and cosmology when no other independent means of verification is available. "Explanation" is really too strong a word for the ability of grand unification to explain the cosmic preponderance of matter. In the language of science, to explain a problem means to account for an observation or phenomenon without recourse to arbitrariness. Quantum mechanics explains the structure of the hydrogen atom, and the wavelengths of the spectral lines it produces; nuclear physics, applied to the first three minutes of the universe, explains the abundance of helium, deuterium, and lithium. But grand unified theories do not in this strict sense explain the matter-antimatter asymmetry of the universe. The best that can be said is that grand unification contains within it the possibility of an explanation, and that specific grand unified theories, suitably adjusted, can be made to produce the desired answer. But the adjustment makes the theory a little less elegant than it was, and the inelegance is tolerated for the sole reason that it is, cosmologically, a boon.

Grand unification would look much better if there were some way of testing it outside the cosmological arena. Physicists certainly recognized this from the outset, and applied themselves to producing practicable tests. They could not hope to capture the massive X and Y particles directly, as they had captured the W and Z particles of electroweak unification, so an indirect test was the likeliest hope. The one thing that seemed plausibly testable was the prediction of proton decay.

Proton decay occurs when a quark inside the proton changes, through the mediating action of an X or Y particle, into some non-quark particle. Because the X or Y is needed only as an intermediary and not as a final product, the energy required to create it can be, so to speak, "borrowed," as long as it is not borrowed for very long. Occasionally, an X particle can spring into existence, mediate a proton decay, then vanish whence it came. The appearance of the X particle appears to violate the law of energy conservation, but if its appearance is so brief that according to the uncertainty principle it could not have been registered on any measuring device, then it doesn't really count. In the quantum world, the X particle never really existed—or maybe it did exist, but it doesn't matter that it did.

At any rate, grand unification permits protons to decay, but very rarely, and the more massive the X particle, the rarer the decays must be. According to the calculations made by inventors of several versions of grand unification, the lifetime of the proton should be at least a million trillion trillion years; that is, if you have a collection of hydrogen atoms (each a proton with an electron) then after a million trillion trillion years, half of them will have decayed into something else. Given that the universe itself is a mere ten billion years old, it seems foolish to imagine that such a prediction can be practically tested. But small amounts of anything contain lots of atoms. A gram of water contains almost a trillion trillion atoms; in a billion grams of water, which is a thousand tons, there will be a billion trillion trillion atoms, and if grand unification is right, as many as one thousand of these should decay in the course of a year. A thousand tons of water needs a tank a mere ten meters on a side—no bigger than an average swimming pool. We end up with an experiment to test grand unification which is surprisingly domestic in scale. Fill a swimming pool with water, surround it with suitable detectors, and count how many protons decay in a year.

In practical terms, the experiment requires a lot of care. The water has to be distilled water, to avoid interference from traces of radioactive impurities, and the tank has to be placed deep underground, because the detectors that are looking for signals from proton decay would, on the surface of the Earth, be overwhelmed by the rain of energetic muons from cosmic rays. Nevertheless, the essence of the experiment involves no more than setting up a big tank of water in a secluded place, then settling down to watch and wait. Efforts have been made in several places to look for proton decay, the most sophisticated experiments being one in a former zinc mine at Kamioka, north of Tokyo, and another in an old Morton Salt mine underneath Lake Erie, not far from Cleveland.

Proton-decay experiments have now logged more than a decade of operation, but not a single incontrovertible proton decay has been seen (the experiments are such that proton-decay candidates turn up fairly frequently, but each must be carefully examined to determine whether it is a genuine decay or, as is always more likely, an event caused by natural radioactivity or by a cosmic ray of unusual energy making it all the way from the Earth's upper atmosphere down to the tank of water thousands of feet below ground). The longer the physicists observed and saw no protons decaying, the longer the proton lifetime had to be. After some years of operation, it was evident that the simplest predictions of grand unification could not be right, otherwise proton decay would have been detected. Needless to say, theoretical physicists have been able to adjust their theories without too much trouble so as to make the proton lifetime come out a little longer.

The ambitious underground proton-decay experiments have been something of a disappointment, in that they may produce direct verification of one of the most striking predictions of grand unification but haven't so far. But the lack of a direct verification hasn't delivered a mortal blow to grand unification, because the theories have been altered to accommodate the news from the water tanks.

More pertinently, as far as cosmology is concerned, the detection of proton decay will not by itself prove that grand unification is responsible for generating the cosmological preference for matter over antimatter. The crucial ingredient in the cosmological argument is the one-part-in-a-billion asymmetry between baryon-creating reactions and antibaryon-creating reactions. It is not surprising to discover that this same asymmetry translates into a tiny differential between the decay rate of protons and antiprotons: in the time needed for a billion protons to turn into other particles, a billion and one antiprotons ought to decay. But if a decade of experiments has yet to yield one certain detection of proton decay, it would require at least ten billion years of experimentation, or one year with a water tank a billion times bigger, to pin down the proton decay rate with sufficient precision to test the asymmetry—assuming, of course, that a similar experiment involving an antitank of antiwater surrounded by antidetectors has been running in parallel in order to measure the antiproton decay rate with the same accuracy. This sort of experiment is, sadly, as unfeasible as the sort of particle accelerator that would create X and Y particles and test grand unification directly.

For particle physicists and cosmologists alike, grand unification is in limbo. There is a consensus that something along these general lines must be true, but no prospect of doing experiments to check it in even the barest detail. Particle physicists still do not know which of the many possible implementations of grand unification is the one that rules our world, and they are not likely to find out quickly, but this has not stopped cosmologists from making grand unification, and its supposed explanation of matter-antimatter asymmetry, into part of the lore of their subject. Grand unification, regarded as a general sentiment rather than a specific physical theory, is taken to have solved the problem of where matter in the universe comes from, even though an asymmetry has to be put into the theory and carefully adjusted for no other reason than to produce the desired answer. Grand unification, which is in physical terms an entirely speculative and wholly unverified theory, is nevertheless taken by cosmologists as a done deal, something they can teach to their graduate students. It does not seem to matter that no specific version of the theory has been settled on—one version or another ought to do the job, it seems, and that is good enough for the cosmologists.

Undaunted by this lack of empirical support, cosmologists have gone on to make even more extravagant use of grand unified theories. If the presence of matter rather than antimatter was a puzzle in cosmology, the simple fact that the universe appears to be expanding at the same rate in all directions is an even more basic mystery. When general relativity was first used to make models of the cosmos, it was natural to look at the simplest cases first, and that meant concentrating on uniform models, in which the expansion was defined to be the same everywhere. The real universe looks persuasively like one of these simple cases, which was at first a very satisfactory outcome. But then it became a problem. Mathematicians exploring the subtleties of general relativity soon discovered that they could make theoretical universes behave quite differently, expanding at different rates in different directions, or expanding in some directions and contracting in others, so that the universe might at different times look like a cigar or a football or a discus. Our universe appears to be the simplest of all possible universes allowed by general relativity, and there was no evident reason for this. The uniformity of the universe had to be ascribed, like the matter-antimatter asymmetry before it, to initial conditions; the universe is uniform now because it was uniform at the outset.

On closer examination, a more subtle oddity showed up. Even if it is presumed that some cosmic law makes the universe uniform, so that it can only be closed, flat, or open, it is hard to see how our universe could, after some ten billion years of expansion, still be near the borderline between the closed and open cases. Our universe, judging from estimates of its overall density and rate of expansion, is fairly close to the dividing line between universes that expand forever and those that ultimately collapse back, and may even be precisely at that division. This is a surprising and mysterious fact. If one were constructing universes on paper, starting them off at some early moment with a size and density chosen at random, it would take an enormous amount of luck to produce a flat, or nearly flat, universe. If there is too much matter in the universe, it will collapse back, and if there is too little it will fly apart at great speed. Out of the infinite range of possible densities, the flat universe is a unique case.

More important, as these theoretical universes grow older, they diverge increasingly fast from the special flat case. If the universe has a little more than the critical density, it soon falls behind the growth rate of a flat universe, and if it is a little less dense, it expands faster than a flat universe, and takes off altogether, expanding so fast that any matter would be pulled into tenuousness much too fast for stars or galaxies to have a chance of forming and growing. A flat universe is like a pencil balanced on its point; it might stay upright for a fraction of a second, but to make

it stay upright for about ten billion years is to all intents and purposes impossible.

One way to finesse this difficulty is to say, as always, that the
universe is flat now because it was flat from the start. This solves
the flatness problem only by means of an arbitrary declaration of
the initial properties of the universe. It would be preferable if an
exactly flat universe were required by some physical law, and in
1973 Edward Tryon, of the City University of New York, published a brief argument to the effect that a flat universe is unique
in having zero energy overall.[1] According to his argument, an
open universe, in which expansion beats out the gravitational
attraction of the matter in it, has positive energy, and a closed
universe, in which gravity overcomes expansion and eventually
pulls the matter back again, has negative energy, but in the flat
universe expansion and gravitation are in perfect balance. Tryon
for the most part was restating the original problem in different
language, but identifying the zero-energy content as a special
characteristic of the flat universe made it seem physically more
natural, as if some universe-creation process might more plausibly
give rise to a zero-energy universe than any other. Nonetheless,
without some hint of a mechanism able to turn out one sort of
universe rather than another, Tryon's argument remained no
more than an appealing way of looking at an old puzzle.

But in 1979 Alan Guth, a young particle physicist with no great
achievements to his name, came up with a way of bringing about,
through a real physical process, Tryon's piece of wishful thinking.[2]
Guth called his trick "inflation." In the barest terms, it amounts to
a period of runaway expansion in the early moments of cosmic
history, and its curious effect is to make the universe converge
toward flatness, regardless of how it starts out, rather than diverge
from it. Inflation could, Guth hoped, make any kind of universe
into a flat universe, and so do away with the disagreeable neces-
sity of ascribing flatness to initial conditions of unknown provenance, or to luck. Rarely has cosmology been transformed so
thoroughly by a single idea as it has by inflation. Whole confer-
ences are devoted to the theory of inflation; hardly a single paper
in theoretical cosmology fails to mention it. And yet we may

never be able to decide whether, a fraction of a second into the big bang, inflation did or did not occur.

To understand inflation, we need to know in more detail how unification in particle physics is accomplished. In the case of the electroweak interaction, we have met the photon, which takes care of the electromagnetic part of the unified force, and the W and Z, which have to do with the weak part. The reason that the weak and electromagnetic interactions look different is that the photon has no mass, whereas the W and the Z are massive, and short-lived to boot. To all outward appearances, the photon is an entirely different creature than the W and the Z, and the trick to unification is to find a way of making these particles into siblings. The explanation, according to the electroweak theory of Glashow, Salam, and Weinberg, is that the W and the Z are not intrinsically massive particles but massless particles that have acquired mass by accident. If they had not acquired mass, they would look like photons, but they did acquire mass, and the weak interaction therefore looks different from the electromagnetic interaction. The acquisition of mass by the W and Z breaks apart what would otherwise have been a perfect symmetry between them and the photon, and this notion of symmetry breaking has become central to all attempts to unify the interactions.

There are, in the world as we see it, three interactions between fundamental particles—strong, weak, and electromagnetic. These forces act on different particles, and have different strengths; moreover, electromagnetism obeys an inverse-square law with distance, but the weak force has a much more limited range, while the strong force gets stronger as the distance between quarks increases. To create a unified theory of the interactions means to create a theoretical structure in which these very different forces emerge somehow from a single underlying device. This unified theory ought, moreover, to be wrapped up in a way that appeals to theoretical physicists. It is no good simply shoehorning the forces together and claiming that unification has been achieved. What is wanted is an underlying device with some elegance or symmetry to it, and then some mechanism that disguises the symmetry and generates from it a set of different-looking forces. This is what symmetry breaking is all about: in the interior of a unified theory is a system in which all forces are equal or symmetrical in some way, but the symmetry is then broken, so that the outward appearance of the forces becomes different.

The idea of symmetry breaking is not hard to imagine. Think of a marble rolling about inside an ordinary round bowl. If it is placed at the center of the bowl, the lowest point, it will remain there; if it is given a little tap in one direction or another, it will roll backward and forward on either side of the center of the bowl. The bowl is perfectly symmetric around its center, and so too is any motion of the marble.

Now imagine a more avant-garde kind of bowl, raised at the center, with a circular depression running around the center. A marble placed in it will roll to the lowest spot it can find, which is now not a single point but anywhere along the circle around the center. The marble, resting somewhere along this circle, can roll back and forth across the depression, or it can roll around the bowl along the bottom of the depression; any motion of the marble can be represented as a combination of these two different motions. The important point is that although the shapes of the two bowls are equally symmetric about their centers, the marbles in them behave very differently. In the first case, the marble rolls in the same sort of way no matter in which direction it is pushed; in the second case, there is a combination of confined motions across the depression and free motions along it. The motion of the marble in the second case does not display the full symmetry of the bowl it is rolling around in. The full symmetry of the bowl, as measured by the motion of the marble within it, has been broken, because the resting position of the marble is off center.

In the first case, any motion of the marble around the center of the bowl can be represented as the sum of oscillations back and forth in two perpendicular directions—north-south and east-west, for example. The system, therefore, contains two fundamental oscillations, which are identical in form. In the second case, there are two fundamentally different motions: one an oscillation back and forth across the valley, the other a rolling along the valley floor. If we think of these two bowls as quantum-mechanical systems, then we can think of the oscillations of the marble as the fundamental physical objects—the elementary particles, if you like. The first case, with symmetry preserved, has two identical particles in it, but the second, in which the symmetry is broken, has two distinct particles.

This, in essence, is how symmetry breaking in particle physics is made to work. The marble represents the quantum-mechanical state of the system, and the oscillations of the marble represent the fundamental particles in the system. When the marble is motionless, there are no oscillations, therefore no particles: this is a vacuum. When the system has some energy, the marble moves around, and its oscillations can be thought of as a set of particles. In the first system, all the particles are identical, but in the second there are two distinct varieties. The tricky part in all this is understanding what the shape of the bowl represents. It is, just as with a real marble in a real bowl, something that defines the dynamics of the system as a whole, but in the particle-physics version of all this we can think of the marble and the bowl as existing only in some abstract mathematical way. The truth is that this method of breaking symmetry in particle physics—it is called the Higgs mechanism, after the Scottish physicist Peter Higgs, who established how it could be used in quantum-mechanical theories— may as well be regarded as mathematical invention. The shape of the bowl is something that the designer of a specific theory adjusts by hand to make the particles come out with the desired properties. The Higgs mechanism is neither intuitive nor obvious; as with so many things in modern physics, one simply gets used it. Its popularity is due to what it does: it allows a system with a basic symmetry to manifest itself in both symmetric and asymmetric ways, depending on what kind of bowl shape is chosen. The simple bowl provides two identical particles, while the bowl with the raised center provides two different particles, both of which nevertheless come from the same system.

This is exactly what unification is all about: the photon and the W and the Z particles all come from the same system, but they are not identical. In this instance, there are four particles rather than two, so the bowl we have been imagining is too simple for this particular job. As unification theories are made more elaborate, to include more particles and more interactions, physicists are forced to deal with more complicated bowls, existing in more than three dimensions (abstract, mathematical dimensions, that is). Nevertheless, the principle remains the same.

The Higgs mechanism, as a way of explaining different particles in terms of the fundamental oscillations of a single quantum-mechanical system, is acknowledged to be an ingenious piece of physics. Whether it is attractive is another matter. It has, when displayed in all its mathematical splendor, a certain flamboyant appeal, like some wild piece of modern architecture; but one cannot help wishing the whole thing were done more simply. Sheldon Glashow, one of the three inventors of the unified electroweak theory and therefore a user of the Higgs mechanism, has referred to it as the toilet in the house of particle physics—the "necessary room," as the Victorians called it.[3] Necessary, but not spoken of.

The biggest objection to the Higgs mechanism is that it was invented to accomplish a particular end, and is designed for that purpose and no other. It works—in theory, at least—but introduces several more arbitrary features into what was supposed to be a simplifying physical theory. A new particle called the Higgs boson turns up any time the Higgs mechanism is used to break a symmetry in particle physics, and a compelling reason to build the Superconducting Supercollider and the Large Hadron Collider is to smash protons into each other with enough energy to create a few Higgs bosons from the debris. Finding them would be the final proof that electroweak unification has been correctly done, and it will demonstrate that the Higgs mechanism is what nature uses, inelegant though it may be. (If the these machines don't find the Higgs boson, particle physics will be thrown into a turmoil, but that is another story.)

If the Higgs boson for electroweak unification is found, physicists will be encouraged to suppose that grand unification also uses the Higgs mechanism. There is at least some economy of design here: particle physics will be a house with two toilets, but the toilets will be of the same manufacture. The inescapable difficulty is that, like the X and Y particles, the Higgs boson for grand unification will be far beyond the reach of the supercolliders or any conceivable supersuccessors to them; the energies needed to make the particles that constitute the internal machinery of grand unification are more than a trillion times what the supercolliders will muster. But cosmology can perhaps provide a laboratory to test grand unification, and it was in this context that Alan Guth came up with the idea of inflation.

Let us go back to our marble in a bowl, which we will now think of as the Higgs mechanism for grand unification. The bowl is of the design with a rise in the center, so that the marble's natural resting place is somewhere in the circular depression around the center. When the marble is off center, symmetry is broken and the strong interaction is distinct from the electroweak interaction. But very early in cosmic history, the temperature was so high that the marble had far too much energy to sit quietly at one place in the bowl; it rattled all over the place at high speed. At a sufficiently high temperature the precise shape of the bottom of the bowl is of no significance, because the marble is able to move freely within it. But if the marble moves equally in all directions, symmetry is restored and there is no distinction between the strong and the electroweak interaction. At high enough temperatures in the early universe, there is just one electrostrongweak interaction.

In the simplest sort of Higgs mechanism, the marble would subside as the universe cooled, and would at some point have too little energy to cross easily from one side of the bowl to the other over the central rise. At this point the marble would be confined to the circular depression, its motion would no longer be the same in all directions, and symmetry would have been broken. What Guth realized was that the transition from symmetry to asymmetry need not be smooth. Guth's idea of inflation, in its original form, necessitated a further complication to the shape of the bowl: it should still be raised at the center, but right at the center there must now be a second depression. What can now happen is that as the universe cools, the marble gets stuck in the central depression instead of in the outer one, where it has to be for the symmetry between the strong and electroweak interactions to be broken. The depression at the center is higher than the outer one, but as the universe cools the marble has less energy, and it is unable to hop over the rise in the center to the circular depression below.

What happens next is cosmological inflation. Heat energy, manifested as oscillations of the marble back and forth, diminishes as the universe expands. But there is also an energy associated with the height of the marble above the bottom of the bowl. When the marble is stuck in the central depression, it is higher up than it would be if it were in the outer depression, and this height shows us, as far as cosmology is concerned, as a quantity of energy. But it is a funny kind of energy. Even when the universe has cooled to the point at which the marble is motionless, which means that there is a total absence of heat energy, the height of the marble's resting place remains just as it was: there is a vacuum, and yet there is energy in it. And no matter how much the universe keeps on expanding, the energy of the vacuum stays the same, because the height of the marble remains the same. A vacuum energy is a contradiction in conventional terms: a vacuum is defined as that which has no content and no properties. It is a matter of semantics, to some extent, whether one thinks of this new energy as a strange kind of stuff that does not diminish even as the universe expands or as a vacuum with some energy to it. If a vacuum means a space empty of heat and particles, then this is a vacuum, even though it contains energy. As seems to be happening with increasing frequency, we must accept the consequences of mathematical theories as they are, and learn to live them as best we can. If that means allowing a vacuum to possess an energy, then so be it. We can call it something other than a vacuum if we wish, but the consequences will be the same.

If cosmological evolution proceeds quietly, and the Higgs mechanism flips from the high state to the low state at the appropriate moment, the vacuum energy disappears, because the marble rolls into the outer depression and comes to a state of rest there. But if the universe gets stuck in the higher, central depression, things go rapidly awry. The energy in particles continues to diminish as the universe expands and cools, but the vacuum energy remains. Before long, it is the only thing left. This has a profound effect on the way the universe expands, which is determined by its contents: the more stuff there is in the universe, the faster it expands. Normally, the process is self-braking, because the expansion dilutes the energy content of the universe, which reduces the driving force behind the expansion. But vacuum energy does not diminish with expansion. If, at the moment the vacuum energy takes charge, the universe doubles its size in a trillion trillion trillionth of a second, then it will double its size again in the next trillion trillion trillionth of a second, and the next, and the next, because at the end of each doubling the density of the universe, which is determined solely by the constant vacuum energy, is exactly the same as it was at the start. The rate of expansion remains constant, and the size of the universe keeps on doubling. This is an exponential, runaway expansion.

This runaway process is Guth's inflation, and it has two far-reaching consequences. First, it very quickly blows up a tiny volume of space into a huge region, big enough to encompass many times over the whole of the presently observable universe. The universe therefore looks the same everywhere, because everything in it originally came from the same microscopic region. Second, inflation makes the universe flat, with density close to the critical value. This is a simple consequence of dynamics. Expansion is influenced by curvature as well as by energy content, and in a normal universe, the influence of the particles decreases more rapidly than the influence of curvature; unless the curvature is precisely zero, as it is for a flat universe, it sooner or later takes over. But if the universe is inflating, the vacuum energy remains constant as the influence of curvature decreases, so curvature too is blown away by the exponential expansion. No matter how the universe starts out—with whatever mixture of matter, radiation, and positive or negative curvature—inflation blows into nothingness everything but the vacuum energy, and the universe is made uniform and flat.

This is perhaps too much of a good thing. Even a modest amount of inflation pushes the universe infinitesimally close to flatness, so that inflationary cosmology makes a firm prediction: the universe today should have exactly the critical density dividing closed from open universes. All astronomers agree that the amount of matter in the universe approaches the critical density; whether it is exactly that is more problematic. The observational evidence is confusing and debatable, but most estimates of the cosmic density come up with values from one tenth to one third of the critical density. On the other hand, observational measurements in cosmology are notoriously changeable, and the fact that everyone gets numbers approaching the critical density may be an indication that the idea of inflation is correct but that astronomers have not yet seen everything there is to see in the universe. Moreover, there has always been a certain sentimental leaning toward flat-universe models, buttressed somewhat by loose arguments like Edward Tryon's, so Guth's assertion that the universe must truly be flat to a very high degree of precision met a prepared and generally receptive audience.

Inflation threw into the realm of answerable questions two problems that had seemed beforehand more or less metaphysical. Why is the universe uniform? Why is the universe flat? Before inflation, uniformity and flatness were assumptions, to be plugged into models as initial conditions and examined no further. Inflation transformed these assumptions into predictions. The overwhelming virtue of inflation was that the initial conditions no longer seemed to matter. The pre-inflation universe could be disorderly, turbulent, irregular, but as long as inflation took hold for a while and expanded the universe so ferociously that all vestiges of its previous appearance were erased, then what would come out depended not at all on what went in, and the universe would come to look the way it does as a consequence of inescapable physical processes. Inflation seemed to be another great triumph for the idea of grand unification, which, having accomplished the relatively modest task of explaining how the universe could contain matter but not antimatter, now claimed an explanation for why it was so large and so uniform. Inflation was a wholly unexpected success of the conjunction of particle physics and cosmology. It was not simply that particle physicists were able to explain certain puzzling features of the universe; further than that, they were able to show that some fundamental attributes of the universe—its size and shape—which had previously been considered inexplicable, to be taken for granted, were in fact susceptible to physical explanation.

But this is to overstate the virtues of inflation. The first difficulty, which Guth himself recognized, was that inflation, in its original form, created a universe that was uniform and flat to an infinitely high degree—and completely devoid of matter. This was vacuum energy, after all. Inflation worked precisely because the universe got stuck in the wrong state, with symmetry unbroken and a non-zero vacuum energy. To get any benefit from Guth's innovation, we cannot leave the universe suspended high and dry, perfectly empty and expanding crazily, all matter, radiation, and curvature having been swept away. From this desolation a habitable universe must be retrieved; the marble must somehow move from the "wrong" vacuum, with its non-zero energy, to the "right" vacuum, around the edge of the bowl, where symmetry is broken and there is no vacuum energy. But the return to normalcy is not easily done.

In strict classical terms, if the universe gets stuck in the wrong vacuum, it gets stuck forever. If it does not have enough energy to get out at the beginning, it certainly will not have enough energy later on, because inflation dries up all sources of energy except the vacuum itself. Quantum physics, fortunately, offers a way around—or rather, through—this obstacle. To quantum particles, no barrier is impenetrable; a classical particle has to go over a barrier, and can do so only if it has enough energy, but a quantum particle can go through a barrier, even when its energy is too small to take it over the top. This phenomenon, known as tunneling, looks peculiar but is not hard to understand. Whereas a classical particle confined in a depression is truly confined, the uncertainty principle dictates that a quantum particle cannot, with absolute classical certainty, be located inside the valley. The marble we are using to denote the state of the universe is a quantum marble, described not by a strict location but by a quantum-mechanical wavefunction that spreads throughout space. There is always some probability—small, but not zero—that the marble in the central depression will spontaneously find itself beyond the rise that confines it.

The tunneling phenomenon allows the quantum marble, with some small probability, to materialize on the outside of the central depression; it can tunnel, in other words, from the wrong vacuum to the right vacuum, bringing inflation to an end. But in Guth's original inflationary universe, there was an inescapable dilemma.

The probability of tunneling decreases as the height of the barrier between the wrong and right vacuums is made greater. To make inflation last long enough to produce a sufficiently flat and uniform universe, the barrier between the wrong and the right vacuum states had to be of a certain minimum height, so that tunneling did not occur too easily. But if the barrier were made high enough for inflation to work, tunneling would be so infrequent that the whole universe could never move into the right vacuum: bits and pieces of it would tunnel through from time to time, but only as sporadic, widely separated "bubbles" of the right vacuum embedded in the still-inflating universe of the wrong vacuum.

This was no use at all. The universe would emerge from the inflationary stage, if it emerged at all, as a series of disconnected empty bubbles; all the good work done by inflation in creating a uniform universe is lost when the inflationary era ends so chaotically. Still, inflation was such a compelling idea that these practical difficulties did not persuade physicists to throw the whole concept aside. In only a year or two, a revised version of the theory, invented simultaneously by Paul Steinhardt and Andreas Albrecht in the United States and Andrei Linde in Moscow, was cooked up. In new inflation, as this version became known, the barrier between the wrong and the right vacuum states was taken away, so that the universe could roll down unimpeded from one to the other. This seems like an odd thing to do, because getting stuck in the wrong state is what makes inflation work, but Steinhardt, Albrecht, and Linde had an ingenious thought: even if the barrier is taken away, the universe will not move instantaneously from the wrong to the right vacuum state, because it takes a certain amount of time to roll from one to the other. And while the universe is rolling, it is still expanding in runaway fashion, because the vacuum energy is still there.

In new inflation, the shape of the bowl is back to what we originally imagined: a central rise surrounded by a circular depression. As inflation begins, the marble sits at the middle, and even though it starts rolling immediately toward the circular depression—the right vacuum—it takes a while to get there. The essential element in new inflation is to insure that this transit time is long enough to yield the benefits of runaway expansion, which means in practice making sure that the center of the bowl is not too steeply inclined. The marble then rolls only slowly toward the right vacuum, allowing inflation enough time to make the universe smooth and flat, and when it reaches the right vacuum it rolls quietly and smoothly into it, guaranteeing an uneventful switch to a normal, non-inflating universe.

But this didn't necessarily work, either. During inflation, matter and radiation have been diluted away to nothing, and only the vacuum energy remains. Somehow, this energy has to be transformed back into ordinary particles (quarks, leptons, photons, neutrinos) to fill the inflated, smoothed-out universe with all the stuff needed to make a three-degree microwave background, to manufacture the desired amounts of helium and deuterium and lithium, and to provide the material of stars and galaxies. Most important, the universe after inflation must get sufficiently hot to create at least some of the X and Y particles of grand unification, in order for the cosmic baryon number to be generated. It is no good using grand unification to accomplish inflation if in the process the original benefit, the explanation of why there is matter but no antimatter in the universe, is lost.

To make sure that the universe after inflation is sufficiently hot, the transformation of vacuum energy into particles must happen abruptly; if it is a slow process—a gradual transformation going on while the universe continues to expand and cool—the maximum temperature attained will be too small for baryons to be made by the X and Y particles. On the one hand, the transformation back to a normal universe must be smooth, so that the benefits of new inflation are not lost; on the other hand, the transformation must be rapid, so that there is enough reheating to make baryons. These two demands can be reconciled, but only just. What is now needed, in terms of the shape of the bowl, is a large, very shallow rise in the center, across which the marble can roll with sufficient stateliness to let inflation do its job, ringed by a steepsided ditch, into which the marble falls abruptly and violently, turning all the energy of the vacuum into rapid oscillations and thence into all the required particles that constitute the stuff of the universe.

There is one final problem, which applies to all versions of inflation. We have taken it for granted so far that when the marble falls at last into the right vacuum, all vacuum energy disappears. This need not be so, by any means. As far as the workings of particle physics are concerned, the only thing that matters is that the vacuum corresponding to the broken symmetry be lower in energy than the vacuum corresponding to the unbroken symmetry, so that the state of broken symmetry is the preferred state of the theory. But for inflationary cosmology to produce the right results, the final resting place of the marble must have zero vacuum energy, because we know that the universe is not now undergoing exponential expansion. This means that the bottom of the ditch into which the marble eventually falls must be close to the absolute zero of vacuum energy, even though there is no reason in particle physics for it to have any special value of vacuum energy, positive or negative. In some versions of grand unification, it is much more natural to expect the vacuum in the center (the wrong vacuum) to be at zero absolute energy, with the right vacuum sitting lower down and having a negative vacuum energy. For cosmological purposes, this would be a disaster, so, the particle physicists conclude, such theories must not be correct.

The idea behind inflation, let us recall, was that the workings of the Higgs mechanism would transform any sort of bizarre, irregular, non-uniform big bang into a large and smoothly expanding universe, just as we see. This would do away with any need for the universe to start out in a special way; anything could be put in, and inflation would insure that the right kind of universe would come out. There would be no need to "fine-tune" the initial conditions so as to make the universe we observe; the universe we live in would be created naturally for us by inflation.

But the simplest version of inflation, as originally conceived by Guth, did not work; it inflated happily, but could not generate from the ballooning vacuum energy a pleasantly uniform universe of matter and radiation. New inflation promised to generate a genuinely smooth universe, but that too was not automatic, and more adjustments had to be made to insure that the reheated universe, after inflation was over with, was sufficiently hot to produce the requisite excess of matter over antimatter. And finally, one last adjustment had to be made to insure that the successfully reheated universe did not itself contain any significant amount of vacuum energy.

All these things can be done, so perhaps the physicists and cosmologists can say that inflation is a success. But how has this success been bought? The Higgs mechanism—the shape of the bowl across whose floor the marble rolls—has to be adjusted with fantastic precision. There must be a wide, almost-but-not-quite flat central area ringed by a ditch about ten thousand times as deep as it is wide, and the bottom of this crevice must with a very high degree of precision sit at the absolute zero of vacuum energy. None of these things happen of their own accord, and must be suitably arranged by the particle physicists who are building grand unification, but their only reason for arranging everything so minutely is to make their cosmologist colleagues happy. In short, the need for fine-tuning the initial conditions of the universe has been obviated by fine-tuning the Higgs mechanism instead.

In this sense, inflation is not the greatest triumph of particle physics in cosmology but its greatest misapplication. Originally, particle physicists contributed to cosmology by providing theories that could push back understanding of the universe to earlier times. Later, particle physicists found that they could test half-formed theoretical ideas by applying them to cosmology and seeing if plausible results emerged. But with inflation, particle physicists have begun to design theories whose sole purpose is not to solve a problem in particle physics but to make cosmologists happy. Inflation is a nice idea; it would be pleasing if particle physics worked in such a way that it made the universe large and uniform. But there is no substantial evidence that inflation actually occurred, and its biggest prediction, that the universe should be exactly flat, may not even be true. Nevertheless, so enamored are particle physicists of the idea of inflation as a cosmic panacea that they have taken to inventing theories that do nothing except make inflation work. The argument is circular—cosmologists like inflation because particle physicists can provide it, and particle physicists provide it because cosmologists like it—and has proved, so far, immune to test.

7

Dark Secrets

THE ONE INESCAPABLE prediction of inflation is that the universe must have exactly the critical density: it must be poised at the borderline between universes that expand forever and those that one day begin to fall in on themselves. But according to the astronomers the amount of visible material in the universe today makes up no more than a few percent of the critical density, in which case the universe is open and will expand forever. Inflation nevertheless remains a very popular theory. How can this be?

For one thing, cosmological observations have been notoriously unreliable. Edwin Hubble's first measurement of the expansion rate of the universe was about ten times too high, which made the deduced age of the universe ten times too small. This was one of the reasons the steady-state cosmology prospered briefly, but in due course the Hubble constant was revised downward and the age problem vanished. One lesson, therefore, is that anyone who proposes a novel cosmology because of a single apparent discrepancy between theory and observation is likely to be disappointed; discrepancies sometimes vanish on closer examination. Equally, cosmological models that seem to contradict the observational evidence should not necessarily be dismissed, because the observations may well change. Hermann Bondi, one of the founders of the steady-state theory, once declared that one should never believe an observation in cosmology until it is supported by theory.[1] Proponents of inflation have grounds for thinking that the cosmic density really is at the critical value, and that it is the observers who are missing something.

There are, indeed, longstanding reasons to believe that there is more matter in the universe than the eye beholds. Back in the 1920s, the Dutch astronomer Jan Oort undertook an astronomical accounting exercise to see what our immediate galactic neighborhood contained. We live in a typical spiral galaxy, consisting of a bright central bulge embedded in a flat circular disk, and our Sun is about two-thirds of the way to the edge. Oort wanted to know how thick the disk of the galaxy was at the Sun's position. By carefully observing and cataloging stars, he was able to map out our neighborhood, and he concluded that the galactic disk in our vicinity is two or three thousand light-years deep; by comparison, the diameter of the galaxy is close to one hundred thousand light-years.

Oort was then able to make a calculation. What keeps the galactic disk thin is the gravity of the disk itself. Any star moving away from the plane of the galaxy will be pulled back by gravity, just as a ball thrown into the air is pulled back to Earth. Knowing the typical speeds of stars relative to the Sun, Oort could work out how much mass must be in the disk to maintain it at its measured thickness. At the same time, he knew how many stars were in his catalog, and could add them up to get a direct measure of the mass of the disk, or at least that portion of the mass that was in the form of stars. To his surprise, Oort's first estimate of the mass was about the twice the second. The gravity of the galactic disk, in other words, indicated the presence of about twice as much mass as could be directly seen.

Oort's analysis has been repeated and updated many times, but the basic conclusion remains the same: there is more mass out there than is visible to astronomers. The discrepancy became known as the problem of the "missing mass," although this is really a misnomer: the mass is there, but we can't see it. In recent years, astronomers have taken to calling it the "missing light" problem instead, and referring to the stuff that can't be seen, but which betrays itself by its gravitational effects, as "dark matter."

Oort's conclusion is not as puzzling as it might seem. Oort himself knew that faint stars outnumber bright ones, and it seemed possible that the dark matter might be nothing more mysterious than a lot of small stars, much less massive than the Sun and thus too faint too be seen.

As astronomers cast their nets wider, however, dark matter kept turning up in greater abundance. From his studies in the 1940s of the motions of galaxies within small groups and clusters of galaxies, the Swiss-American astronomer Fritz Zwicky concluded that there must be almost ten times as much matter in a typical group of galaxies as would be inferred from the brightness of the galaxies alone. But Zwicky's was a lone voice; most astronomers ignored his claims, deciding that the dynamics of galaxy clusters were not sufficiently understood to make credible so strange a conclusion. Not until the early 1970s did dark matter emerge from the theoretical netherworld to become a staple of modern cosmology.   Vera Rubin and her colleagues at the Carnegie Institution in Washington undertook a survey of normal spiral galaxies with the aim of obtaining accurate "rotation curves"—graphs plotting the velocity at which objects orbit around the center of a galaxy against their distance from the center. They wanted to measure rotation curves right out to the edges of galaxies, which would enable them, by a simple application of the law of gravity, to calculate the masses of the galaxies. The problem, of course, is that at the faint outer edge of a galaxy it is hard to find anything you can measure the velocity of. Rubin solved this problem by tracking the motions of faint clouds of hydrogen gas, which exist in many cases well beyond what a casual observer would call the edge of the galaxy. By observing these tracers, Rubin hoped to be able to measure accurately total galactic masses.

But there was a surprise. If the hydrogen clouds had in fact been beyond the edge of the galaxy, their speeds of rotation should have decreased with distance in a predictable way. Instead the rotation curves held steady. The gas clouds orbited at about the same velocity, no matter how far away they were from the center. Even though there was nothing to see, the dark space between the gas clouds and the body of the galaxy must therefore contain substantial quantities of mass. It seemed that what we call a galaxy is but the visible core of a much larger but dark object with, as Zwicky had said, a total mass typically ten times what the astronomer counting only the bright material would measure. All galaxies, astronomers now believe, sit in a much more massive "halo" of dark matter.

But such observations did not reveal what the dark matter might be. Obviously it could not be the normal stars or the hot gas and dust that make up the visible parts of galaxies; it might be tiny dim stars, or planets, or rocks; it might as well be baseballs, or watermelons. The only way to decide among the virtually limitless options was to think about where the dark matter could have come from. One restriction was suggested by the success of cosmological nucleosynthesis, which accounts for the observed abundances of helium, deuterium, and lithium only if "normal" matter—meaning anything made of protons, neutrons, and electrons—constitutes 10 percent or less of the critical density of the universe. The best present estimates suggest that if each galaxy is given a complement of dark matter ten times as great as its visible content, then the density of the universe overall is about 10 to 20 percent of the critical density. It is possible, if this is correct, that the universe is made entirely of normal matter, some of it visible, most of it not, but in this case inflation cannot have happened, because the total density is well below the critical value.

The observational evidence has grown more complicated in recent years. Astronomers have been able to map out the galaxies in large regions of the universe, and, like Zwicky, they can use their maps and catalogs, coupled with an understanding of gravitational dynamics, to weigh these regions. A general trend emerges: the bigger the volume sampled, the greater the apparent ratio of dark matter to visible matter. Individual galaxies and their dark-matter burdens might add up to about 10 percent of the critical density, but when clusters and superclusters are added in the figure rises as high as 50 percent or 70 percent—closer to 100 percent of the critical density, which inflation demands, but never quite getting there. Still, for those who like inflation, the trend is encouraging. It begins to seem plausible that if cosmologists could weigh the whole universe at once, rather than bits of it at a time, a total density equal to the critical value might appear.

Whether this is wishful thinking or a reasonable extrapolation

from established trends is hard to say, but the inflationary prediction of an exactly critical density looks closer to observational
reality than it did a decade ago, when inflation was invented.
Now a new problem crops up, however. Cosmological nucleosynthesis limits the density of normal matter—protons, neutrons, and
electrons—to no more than 10 percent of the critical density, if
the abundances of helium, deuterium, and lithium are to come
out right. So if the true density, including dark matter, is in fact
equal to the critical density, then 90 percent of the stuff in the
universe must be made out of something other than protons,
neutrons, and electrons. The dark matter is not just dark; it is
mysterious. It is not enough to imagine that the universe is full of
ordinary matter that for some reason is not glowing, so that we
cannot see it. The dark matter is something wholly new, a novel
form of matter not so far seen in physics laboratories; its presence
is deduced rather than strictly proved, but the deduction is at its
simplest no more than an application of the law of gravity to
uncontroversial astronomical observations.

This is a step into the dark in more ways than one. Cosmological theorizing, if done with the aim of constructing a simple
model of the universe, neither predicts dark matter nor requires it
to exist, but observations indicate that it is there. Theories must be
made more complicated to accommodate this new fact, but as
they become more complicated they inevitably move away from
the guiding principle of simplicity that has always played such a
large role in cosmologists' efforts to choose between theories. And
since the dark matter, by its very nature, is undetectable by the
usual astronomical means, the empirical grounds for choosing
between theories become weaker.

Particle physics contributed to the dark-matter problem, by providing inflationary cosmological models that require the cosmic
density to take on the critical value, but it may also help to
resolve the problem. Theories of particle physics are, after all, full
of particles, and some new type of particle, gathering unseen in
the halos around galaxies, may fill the role of dark matter.

Alan Guth based inflation on the symmetry breaking inherent
in grand unified theories linking the strong and electroweak

interactions. But Guth's idea didn't work very well, and particle physicists were encouraged to look to other physics theories for ways of making inflation work smoothly. Such theories, during the 1980s, were beginning to abound, because particle physicists were so emboldened by their successes that they started to work on theories to explain problems that until recently had not looked like problems at all. Such a venture yielded theories based on something called "supersymmetry," and supersymmetric theories came with all sorts of capacities and flexibilities for the cosmologists to turn to their advantage.

Supersymmetry goes a step beyond attempts to unify the forces acting on elementary particles. If the strong, weak, and electromagnetic forces are brought under one roof, it will not mean that particle physics has been all tidied up; other curious facts about the elementary particles come to the fore. Most notable is their division into two sorts: those that transmit the forces and those on which the forces act. The force-transmitters—the photon, the W and Z particles of electroweak unification, the gluons that tie quarks together, the supposed X and Y particles of grand unification—all possess quantum-mechanical spin in whole-number units (the photon has one unit of spin, for example). But the elementary particles on which these forces act—the electron and its cousins the muon and tauon, the neutrinos, and the quarks—all have spin that comes in half units (the electron has half a unit of spin). This is the kind of obvious and unavoidable empirical fact that, in the mind of the theoretical physicist, demands explanation; it is unacceptable to suppose that so striking a division can be mere accident.

Moreover, there is a fundamental difference between the properties of these two types of particle. Technically, particles with whole-number spin are called Bose-Einstein particles (bosons, for short), while those with half-number spin are Fermi-Dirac particles (fermions). Under the rules of quantum mechanics, fermions obey Pauli's famous exclusion principle, which states that no two particles with identical properties and attributes can occupy the same quantum-mechanical state. Pauli's principle means that electrons in an atom do not all fall into the lowest possible orbit but successively fill orbits of increasing energy. Since all of chemistry and atomic physics depends on the way that electron orbits arrange themselves around the parent nuclei, the fact that electrons are fermions and therefore obey the exclusion principle is responsible for the way the everyday world is constructed. Similarly, quarks are fermions and obey the exclusion principle, and this insures that quarks build up protons and neutrons and all the exotic baryonic particles of the 1950s and 60s in an orderly fashion. Without the exclusion principle, the strong interaction between quarks would draw them all together into vast globs of quark stuff, the size of stars and bigger. The fact that all the "material," as distinct from force-carrying, elementary particles, belong to the class of fermions and obey the Pauli exclusion principle is responsible for the structure of the world as we know it.

The division of the elementary world into fermions and bosons is therefore a matter of deep significance, but it is entirely unexplained in the standard picture of particle physics, with or without grand unification. This does not sit well with physicists, who once they perceive a profound empirical fact are obliged to begin the task of explaining it. For a long time, this particular fact seemed as unbreachable as it was obvious, but the success of electroweak unification and the presumed success of grand unification encouraged physicists to see all kinds of distinctions in the observable world as the consequence not of fundamental differences but of underlying symmetries that are broken, and therefore concealed from our direct appreciation, by some intervening physical process.

So, inevitably, arose the idea that the grouping of elementary particles into fermions and bosons should be interpreted as the two halves of an unknown whole, with some mechanism yet to be invented responsible for the obvious differences between the fermionic and the bosonic halves. The innovation that made this vague thought into the basis of a plausible theory was a mathematical formulation by which fermions and bosons could be represented as different aspects of a single entity. The price for this innovation was that extra dimensions of space had to be added to the theoretical world. These extra dimensions are not part of conventional geometry, since we obviously see only three dimensions of space and one of time, but can be thought of as forming an addendum to normal spacetime. Each particle is tagged with an arrow in the appended dimensions; when the arrow is pointing up, the particle is perceived in ordinary space as a fermion, and when it is pointing down the particle is perceived as a boson: switching the arrow up and down turns fermions into bosons and back into fermions again. This concealed connection between the two kinds of particles is supersymmetry.

But physicists had no sooner invented supersymmetry than they had to disguise it. In a precisely supersymmetric world, all particles would come in pairs, identical but for one being a fermion and the other a boson. Corresponding to the half-spin electron, a fermion, there would have to be a zero-spin partner, a boson, with precisely the same charge and mass. There is, we can be sure, no such particle. If there were, it would be able to orbit around atomic nuclei just as the electron does, but because this hypothetical particle, being a boson, would not obey Pauli's exclusion principle, there would be nothing to stop many of them piling down into the lowest energy orbit of the atom. They would displace electrons from atoms, and would destroy everything we know of atomic structure. Because this catastrophe does not happen, the bosonic partner to the electron either does not exist, or must have significantly different properties from what a strict application of supersymmetry implies. The first possibility is straightforward but dull, and would mean that supersymmetry is of no relevance to our world. The second possibility is more pregnant with hope for the physicist who likes the idea of supersymmetry: it implies that supersymmetry may be present, if it is sufficiently well hidden. Just as unified theories of the interactions between particles work by asserting that the strong, weak, and electromagnetic forces are disguised aspects of a single fundamental interaction, so the division of particles into fermions and bosons is to be understood as a flawed implementation of supersymmetry, not as an absence of supersymmetry. This method, in particle physics, is becoming the rule for explaining every seeming distinction: turn it into a perfect symmetry, improperly realized.

But even if supersymmetry is broken, as it must be, the fermions and bosons must still pair up somehow. The bosonic partner to the electron may look very different from the electron itself, but it must nevertheless exist. It would have been agreeable to discover that, for example, the electron's "superpartner" was some already known particle—one of the gluons, perhaps. No such luck. Because exact supersymmetry is broken, we must expect that particles and their superpartners have different masses, but in other respects—charge, strangeness, charm, and so on—they have to be identical. Unfortunately, none of the known particles could be thus teamed up, which meant that for every known particle there must exist an as yet undiscovered superpartner, in order to provide supersymmetric theories with a full deck of cards. In the history of particle physics there are many instances, it is true, when a physicist advocating some new theory has predicted the existence of a new particle. Many of these predictions were borne out: the neutrino, the antielectron, the W and Z particles all existed in the minds of theoretical physicists before they were discovered. But proponents of supersymmetry were obliged to postulate not merely the existence of a new particle, or even a new group of particles; they had to double, at a stroke, the number of elementary particles in the world, and then declare that, by mischance, half of them had not yet been found. And not just some randomly selected half, but precisely the half that were partners to all the known particles.

Still, physics makes progress only by pursuing ideas to their conclusion and declaring them possible or impossible. Supersymmetry, like inflation, was based on an original thought of compelling beauty, and even though the reality might turn out to be less appealing than the conception, it would be folly to abandon it at the first hurdle. There was, in any case, a plausible reason why none of the superpartners had yet been observed. Electroweak unification and grand unification are similar in that they both unify interactions by means of the Higgs mechanism, but they operate at entirely different energies. Electroweak unification has been partly verified by the accelerator at CERN, but a similar verification of grand unification would require an accelerator a hundred trillion times more powerful. It becomes a legitimate and disconcerting question to wonder why one theory of particle physics should contain two versions of the Higgs mechanism operating at such wildly different energies. There was nothing to stop physicists from plugging in the appropriate numbers to make the answers come out right, but to the mind of the physicist, as we have seen, the presence of any large number—in this case, the ratio one hundred trillion to one—is something that ought one day to be explained.

Supersymmetry offers a sort of rationale for this large ratio. When supersymmetry is broken, as it must be, all the superparticles become heavier than their known counterparts by a similar factor. The partners of massless or very light particles, such as the photon or the electron, naturally acquire a certain mass, which must be high enough to explain why the partners have not yet been detected. A convenient idea is to make the mass involved in the breaking of supersymmetry similar to the scale of the electroweak-unification energy. The partners of the photon and the electron then acquire masses similar to the mass of the Higgs boson for electroweak unification, which is just beyond the reach of existing particle accelerators. A corollary of this is that if the next generation of accelerators, the Superconducting Supercollider and CERN's Large Hadron Collider, can find the electroweak Higgs boson, they can perhaps find the lightest of the proposed superparticles too. History suggests, however, that if these superparticles don't turn up, there will be strenuous efforts to save supersymmetry by tinkering with it rather than deciding that the whole thing is a failure.

There is to date no real evidence for supersymmetry, but it is so powerfully appealing an idea that if it is not the foundation for a deeper understanding of particles and forces many physicists would agree with John Ellis of CERN that it would be "a waste of a beautiful theory."[2] (The theory may be beautiful, but its introduction caused the quality of nomenclature in particle physics to sink to new lows. Generically, the superparticles have become abbreviated to "sparticles"; in individual cases, the electron begets the "selectron," the neutrino the "sneutrino," and, worst of all, the whole set of quarks turns into a corresponding set of "squarks." Where the addition of an initial S doesn't work, diminutive endings have been resorted to, producing a "photino" to go with the photon, "gluinos" for the gluons, "Wino" and "Zino" for the W and Z, "Higgsino" for the Higgs boson, ad nauseam.)

Supersymmetry may be, for the time being, no more than a theoretical dream, but that has not prevented cosmologists from using it with enthusiasm. To those who are looking for some particle that might constitute the dark matter of the universe, supersymmetry is an attic stuffed with a superabundance of particles. The photino, for example, should have been made in large quantities in the early universe and could be with us today, invisible and undetectable but carrying enough mass to fill up the cosmos with its unseen presence. It or some other superparticle might be lurking in the halos of galaxies and throughout galaxy clusters, providing the massive counterweight that observers say is there. So far, none of the new particles of supersymmetry seems an especially appropriate candidate for the dark matter, but there is at least plenty of choice.

Nor was this the only use cosmologists could think of for supersymmetry. In the early 1980s, inflation ran into trouble once again, and it appeared that grand unification might not be sufficiently flexible to keep inflationary cosmology alive. But supersymmetry, a more complicated theory, contained more features that could be adjusted.

The new problem for inflation was, remarkably, first recognized, then fought over, and finally disposed of, all in the space of a two-week conference at Cambridge University, in the summer of 1982. The Nuffield meeting on quantum gravity, hosted by the Department of Applied Mathematics and Theoretical Physics (Stephen Hawking's bailiwick), had been intended as a workshop to address a variety of theoretical issues arising from attempts to bring quantum mechanics and general relativity into harmony, but it turned into an intense and competitive race to understand inflation, then only a couple of years old. By the time of the meeting, new inflation had supplanted Guth's old inflation, and the general picture of a period of runaway cosmic expansion, followed by an abrupt and efficient transformation of vacuum energy back into real particles and radiation, was understood and accepted. But this, it turned out, was still not enough. The new difficulty was due to the venerable uncertainty principle of quantum mechanics, which in this case meant that there were bound to be small variations in how inflation had proceeded at different places in the universe.

The particular issue recognized at Cambridge in 1982 was that the uncertainty principle must mean that inflation could not end with perfect uniformity. The transformation of vacuum energy into ordinary matter could not be exactly simultaneous throughout the universe, but would begin with a few patches starting to change while the rest of the universe was still full of vacuum energy. As soon as vacuum energy turns into ordinary matter, inflation at that place ceases, and the runaway expansion slows to a much more modest pace. The places that change first are left behind by the runaway expansion elsewhere, and by the time the whole universe has turned back into ordinary matter, untenable variations in the density of the universe from one place to another have been created.

Three groups at the Cambridge meeting latched on to this problem. The technical calculation of the magnitude of the non-uniformities was far from straightforward, and at first each group came up with a different answer. But after a session of late-night thinking and scribbling, one answer emerged—and it was an unwelcome one. After inflation had ceased everywhere, the density variations in the universe were seen to be as large as they possibly could be, like ten-feet-high waves in a swimming pool ten feet deep. What came out of new inflation was not the tolerable, well-ordered universe it had been designed to produce but, alas, a new kind of anarchy.

The theoretical physicists and cosmologists in Cambridge hardly paused to digest this result before they realized that it could be overcome. All they had to do was fiddle some more with the internal workings of symmetry breaking—with the Higgs mechanism, in other words. They drew up a list of requirements that the Higgs mechanism must meet in order to allow inflation to proceed as requested. There had to be, as before, a runaway expansion followed by the abrupt transition from vacuum energy to ordinary matter, but there was an additional requirement: the runaway expansion could not be too prolonged, or the place-to-place variations called for by the uncertainty principle would grow too large and, at the end of inflation, manifest themselves as excessive density variations.

The participants at the Cambridge meeting truly snatched victory from the jaws of defeat, because by the time the meeting dispersed this looming disaster had been declared a positive triumph. Variations in the density of the universe are, at some level, actually required, because the universe today is filled with galaxies and clusters of galaxies which cannot have arisen from a perfectly smooth universe. Density variations of about one part in ten thousand, created at the end of inflation, were deemed just what was needed to explain the presence of galaxies ten billion years later, and inflation had the additional virtue of creating density variations equally on all scales, from millimeters to megaparsecs. Inflation, lo and behold, was now capable of generating not only a universe large and smooth enough to contain the galaxies and clusters we see, but also the seeds of the galaxies and clusters. All it took was a willingness to adjust the internal workings of particle physics with an extra degree of fineness, and in supersymmetry particle physicists had unwittingly provided a theory with enough knobs and levers to allow the cosmologists to succeed. The particle physicists themselves merely wanted the supersymmetry between fermions and bosons to be broken one way or another, so that the selectron and the photino and all the other superpartners of the known particles were made sufficiently undetectable to explain the fact that they had not been detected. The cosmologists were then left to tinker with the precise details of this symmetry breaking, in order to keep inflation.

Like grand unification, inflationary cosmology has become a widely accepted piece of theorizing while remaining entirely conjectural. With the right kind of theory, it can be made to work. Some cosmologists have found even supersymmetry a bit too restrictive, and have decided that inflation is so important a theory that it deserves its very own Higgs mechanism, divorced from any real or imagined symmetry breaking in particle physics. Thus arises, in some theories, a particle called, almost as an admission

of defeat, an "inflaton," belonging to a Higgs mechanism introduced into physics for no other reason than to make inflation
work.

Albert Einstein once memorably asked whether God had any
choice in the creation of the universe. Were the laws of nature
and of physics such that our kind of universe would happen, in
other words, inevitably and automatically? Inflation began as an
answer to this question. Alan Guth meant to demonstrate that
grand unification would lead to cosmological inflation, which
would in turn lead to a large, uniform, long-lived universe—the
very thing we inhabit. But now this logic has been turned inside
out: to get our universe, so the new argument goes, there must
have been the right kind of inflation at an early time, but neither
grand unification nor (probably) supersymmetry is quite up to the
task of making the right kind of inflation happen, so the true theory of particle physics must contain some other additional piece
that drives inflation. While Einstein wanted to know if the universe had emerged inevitably from the laws of physics, his successors now want to declare that the laws of physics are inescapably
determined by the way the universe looks.

Perhaps this is too dim a view. The universe, with its array of
galaxies and clusters, is a complicated thing to explain, and while
one may hope for a simple theory to explain it all, it is more realistic, perhaps, to expect that the final explanation may be complicated as well. A simple theory that appears to explain in broad
terms some basic part of the structure of the universe acquires
thereby a good deal of credence, and if some more complicated
theory of particle physics, applied to the early moments of cosmic
history, proves capable of explaining all that followed, down to
the last ounce of dark matter and the last detail of the distribution
of galaxies, there will be reason to celebrate. The only real hope
for pinning theories down is, as always in science, to examine as
minutely as possible their consequences. Perhaps some simple
theory will work so beautifully that its truth will seem compelling. The way forward, at any rate, lies in understanding how
the details of the early universe evolve into the details of what we
see now around us.

It is not too difficult to grasp qualitatively how galaxies came into being. Newton had the basic idea right in a letter of 1692 to the theologian Richard Bentley.[3] He surmised that if matter were spread uniformly throughout the universe, it would not stay that way: the tiniest excess of material in one place would pull other material toward it, so a clump of material would grow at the expense of its surroundings. Newton also realized that in an infinite universe many such clumps would begin to grow, and he suggests that these clumps could be the stars. The modern picture is not all that different: cosmologists imagine that small irregularities in the distribution of gas across the early universe grew, through the action of gravity, into clumps that became galaxies. The main difference from what Newton imagined is that stars are thought to be secondary objects, forming only after gas has aggregated into galaxy-sized objects; it is the galaxies that are now thought to be the building blocks of cosmological structure.

This is a sketch of galaxy formation, not a real theory. It does not begin to explain why galaxies fall into different types, or in what manner they are scattered across the sky. The latter issue— the distribution of galaxies across the sky, and their aggregation into clusters and clusters of clusters—has become the central issue of cosmology in the last decade. Galaxies are not scattered across the sky at random. If you scatter a few grains of sand on a sheet of paper, there will be, by statistical accident, some clusters of sand grains and some empty spaces. But the galaxies in the sky form much larger clusters and leave much larger tracts of empty space between them than such a scattering would create. Clusters of galaxies are larger than randomness alone would generate, and there are more of them. There is large-scale structure in the pattern of galaxies, which could not have arisen by slow gravitational aggregation in an initially uniform distribution of matter, as Newton imagined. This is the fundamental principle of cosmological theories of galaxy formation: there must have been, in the initial pattern of density fluctuations that produced galaxies, preexisting structures that constituted the seeds of what we have before us now. Getting from those initial fluctuations to the present arrangement of galaxies is the basic task of galaxy-formation theories.

The transformation of those fluctuations into galaxies and clusters of galaxies depends on the cosmological setting: the density of the universe, the rate at which it is expanding, the proportion of visible to dark matter, the nature of that dark matter, and so on. Galaxy formation must be considered in the context of some chosen cosmological theory, so cosmologists must juggle all the factors at once and hope to arrive at a simultaneous explanation not just of the galaxy distribution but also of the dark-matter content and the cosmic density. This makes the problem more complicated, and more susceptible to theoretical prejudice. A cosmologist enamored of inflation, for instance, will insist that the universe must have exactly the critical density, and will count as unsatisfactory a theory that successfully explains galaxy clustering only if the density is, say, 20 percent of the critical value. For the moment, no theory works satisfactorily in all respects, leaving cosmologists to fiddle with one parameter or another, according not just to the quantitative success of the model but also to their idea of what a pleasing galaxy-formation theory ought to look like.

The basic tool for this work is a large computer, equipped with a program that can follow the motion and gravitational attraction of millions of particles. Even the largest computers and the most efficient programs are unable to provide a perfect representation of the whole universe at once, so shortcuts of one kind or another are essential. A program that treats each particle as a single galaxy cannot deal with a large volume of the universe, which contains billions of galaxies, and such a program therefore misses the largest structures; a program that is to encompass the whole universe and the biggest structures in it must perforce lump thousands of galaxies together as a single particle, and therefore blurs the smaller structures. Nevertheless, with two decades of experience in the art of numerical computation, cosmologists have learned how to piece together large-scale and small-scale simulations so as to test comprehensively an entire galaxy-formation theory.

An early and profound lesson drawn from these numerically simulated cosmologies was that some sort of dark matter seems to be required as much by the theories as by the observations. Ordinary matter, as it comes together under its own gravitational attraction, turns into stars, which then throw off energy into space in the form of heat and light. The continual dissipation of energy tends to make the stars and gas in a galaxy collapse toward its center. This is good, because it explains how galaxies can form, but bad, because it raises the question of how large clusters of galaxies can survive. Large clusters of galaxies made entirely of normal matter would have trouble forming in the first place, and if they did form would tend to fragment into sub-units. But, cosmologists realized, if there were the right sort of dark matter—a sort interacting only inefficiently with the normal matter—it could create structures that would stabilize and sustain large clusters of galaxies without breaking up into pieces.

There was a moment in the early 1980s when it seemed possible that this dark matter had been identified. A few experiments around the world came up with some evidence that the neutrino, in standard physics strictly a massless particle, might actually have a small mass. The mass per neutrino was tiny, but because there are as many neutrinos in the universe at large as there are photons in the three-degree microwave background, even a tiny mass could add up to a lot for cosmology. It was entirely conceivable that there could be about ten times as much neutrino mass as normal mass, in which case the overall density of the universe could reach the critical value. As a form of dark matter, massive neutrinos had some appeal. Neutrinos are known to exist, and giving a previously unsuspected mass to an existing particle is more palatable than inventing a wholly new particle—the hypothetical photino, for example—to act the part of the dark matter. On top of that, the mass suggested by laboratory experiments was about the right value to be cosmologically significant. There were reasons for taking "neutrino cosmology" seriously.

For one thing, it appeared at first that neutrinos as dark matter might explain the large-scale clustering of galaxies in the universe. Neutrinos, if they are perfectly massless, move at the speed of light, and the small hypothetical mass now given to them would not slow the neutrinos down much. Moreover, because neutrinos collide with each other or with particles of normal matter so rarely, they cannot easily lose any of their energy. What this means is that massive neutrinos, unlike normal matter, cannot collapse to form compact structures. They remain fast-moving, so the tendency for them to fall together under their own gravity is weak. This seemed just right: massive neutrinos could fill up the universe to the critical density, and they could form large, loose structures; within those large structures, normal matter could fall together and lose energy, creating the tightly compacted bright objects we see as galaxies. Alas, this simple picture did not survive very long. The uncollapsibility of massive neutrinos was too much of a good thing. The large structures that would form in a neutrino universe would be too large and too loose to correspond to the observed galactic clusters. And it is impossible for neutrinos to be the dark matter in individual galaxies, because they are too fast-moving to be retained by the gravity of single galaxies. Massive neutrinos ended by failing on two counts: they created too much large-scale structure in the universe, and they could not account for galactic dark matter.

The original laboratory evidence that neutrinos might have a small but cosmologically interesting mass has now more or less been discounted. But the brief and glorious effort that went into the construction of neutrino cosmologies was the opening of a Pandora's box. Particle physicists had offered one candidate for the role of dark matter; it had been rejected, but they had plenty more up their sleeves. Supersymmetry could provide all manner of particles. The photino, for example, is likely to be more massive than the neutrino, but much less numerous in the universe, so that its overall contribution to the cosmic density could conceivably be about the same. Because they are more massive, and therefore more slow-moving, photinos are capable of being entrained in the gravity of individual galaxies, thus constituting the required dark matter, and like neutrinos they can also maintain structures on large scales, because, like neutrinos, they do not easily give up their energy to normal matter. If the fast-moving neutrinos were a form of "hot" dark matter, then the slower-moving photinos were merely "warm." They did a better job, but in the end they, too, failed.

What cosmologists decided they needed was something more sluggish yet—something that could be called "cold" dark matter.

Cold dark matter would be able to collapse into galaxy-sized units, but also, because it interacted little with normal matter, could sustain structures on the scale of galactic clusters without falling apart.

Cold dark matter it was, then. But here was a historic moment in the subject. Until this point, cosmologists had asked the particle physicists to supply the dark matter, in the form of one hypothetical particle or another, and had then done their best to make a workable cosmology from it. When the effort failed, they had gone back to the particle physicists for a new kind of dark matter, something a little more recherché. But after a few go-rounds, the cosmologists had become frustrated. They decided that they could tell, from what they knew about the disposition of galaxies in the sky, what sort of dark matter had to be up there, and they went back to the particle physicists, told them what was desired, and waited to see if the particle physicists could find a way of providing it.

They could, of course. There was a particle known as the axion, which had been invented to account for some small details of quantum chromodynamics, the theory governing the interactions of quarks and gluons. The axion could have almost any mass you cared to give it, so it was convenient to give it the kind of mass the cosmologists wanted. And it would be cold. With this backing from the particle physicists, cold dark matter, based on axions, became, in the late 1980s, the most popular cosmological medium. It seemed capable of getting structure right on the small and large scales equally, and had the right properties to constitute the dark matter in individual galaxies. But the cold dark matter used by the cosmologists was something of their own devising, and the fact that it might be made of this particle called the axion was secondary. If the axion fell into disfavor among physicists, or if it could not continue to meet cosmological requirements as the galaxy-formation models were increasingly refined, it could be set aside, and the cosmologists would wait for the physicists to come up with some other invention. Cold dark matter is first and foremost a cosmological idea, and something for particle physicists to explain if they can.

* * *

All this theorizing was for many years built upon one set of observational data—the observed distribution of galaxies. But there
was another set of data that, in principle, could help resolve matters, and as cosmological theorizing became more intricate, the
need for this extra data became more desperate. The necessary
observations, only recently obtained, were measurements of tiny
temperature variations in the microwave background, from place
to place across the sky. The variations have to be there, if the idea
that galaxies and clusters grew from small density irregularities is
correct. In the modern universe, the photons in the microwave
background sail unimpeded through space, but a long time ago,
when the universe was hot and dense, matter and radiation were
in intimate contact. If there were density variations in the matter,
there must have been comparable variations in the number of
photons at any one point. When the universe was less than a year
old, matter and radiation parted company, but because the radiation was once closely tied to the matter it should have retained an
imprint of the density fluctuations obtaining at the time they lost
contact. As a direct consequence, the three-degree microwave
background should be dappled with tiny temperature variations
that reflect the distribution of matter in the universe long ago.

Any successful theory of galaxy formation has to connect the
pattern of background temperature variations to the patterns of
galaxies and clusters. Through the 1970s and 1980s, measurements
of microwave-background fluctuations had been attempted, down
to a sensitivity of about one part in ten thousand, but nothing had
been found. This was alarming: cosmologists concluded from their
computer programs that if the initial-density variations were no
bigger than this, it would have been essentially impossible for
galaxies to grow to their present size in the time available. But
here again, dark matter came to the rescue. Dark matter, whatever its precise form, would have been less closely tied to the radiation in the early universe than the normal matter was, and
would therefore have left behind smaller imprinted temperature
variations. Now the idea was that galaxies would begin to form by
means of the dark matter alone, with the normal matter falling
into the pre-formed protogalactic dark-matter structures only at a
later date.

The searches for the expected temperature variations in the microwave background became increasingly sensitive, but still they came up with nothing. By the time measurements had got down to a sensitivity of a few parts in a hundred thousand, with no result, another theoretical innovation came along. It had been assumed that dark matter and normal matter might go through somewhat different histories to get to the present pattern of galaxies and clusters in the universe, but it was nevertheless assumed that the visible distribution of galaxies reflects exactly the underlying distribution of dark matter. This need not be so, however. It is possible that normal matter turns into galaxies, and becomes visible, not in every place where there is an excess of dark matter but only where there are the very deepest concentrations of it. This means that the galaxy distribution we see, which comprises normal matter alone, is an exaggerated copy of the dark-matter distribution underneath, and this in turn means that the observed galaxy distribution may have arisen from a smaller set of density variations than was otherwise thought. This notion is called "biasing," and it was yet another way to reconcile the obvious presence of large structures in the modern universe with the persistently smooth appearance of the microwave background.

With biasing added, cold dark matter continued to be the theory of choice into the early 1990s. Since then, it has suffered a death followed by a partial resurrection. At the beginning of 1991, a galaxy survey of unprecedented size was published, which appeared to establish conclusively that there was more large-scale structure in the universe than even the most exquisitely adjusted cold-dark-matter theory could account for.[4] The observed galactic clusters were too big and too numerous. The widely publicized "death of cold dark matter" was transformed, in many a popular account, into the death of the big bang itself. This was clearly exaggeration, since cold dark matter is but one part of a much larger picture, but the confusion was perhaps understandable: the cold-dark-matter model had been so widely accepted as the standard for galaxy-formation theories that it had become—somewhat like inflation—a seemingly inseparable part of the big theoretical picture. Even before this apparently mortal blow, however, some skeptics had begun to wonder whether any sort of dark-matter theory could work. The looming danger was that the whole idea of galaxies growing from small initial-density variations might be fundamentally misguided, since no sign of irregularities in the microwave background had been found.

Many of the doubts were thrust aside, however, with the stunning news, in April of 1992, that the long-sought temperature fluctuations had at last been detected. The Cosmic Background Explorer (COBE), a satellite built expressly to study the microwave background, had by then scanned the entire sky several times from its vantage point above the interferences of the Earth's atmosphere. Analysis of the first year's worth of data revealed the presence of minute temperature variations at a level of a few parts per million. There had been, evidently, small density variations in the early universe, and these had, presumably, somehow grown into galaxies and galactic clusters. The form of the fluctuations detected by COBE was declared to be consistent with what inflation predicted, and consistent as well with the presence of cold dark matter, although the measurements were by no means detailed enough to favor one galaxy-formation theory over another. That will have to wait for more results from COBE, which is still in Earth orbit, as well as from other experiments, ground-based and balloon- and rocket-borne, designed to measure other aspects of the microwave background. Nevertheless, the COBE measurements gave cosmologists confidence that they had not been wasting their time all those years. It may still be difficult to find a theory that leads precisely from the temperature variations to the observed pattern of galaxies and clusters, and it looks increasingly unlikely that any such theory will be simple or tidy, but at least there remains the hope that a theory will one day be found.

Cosmologists have now set themselves to accommodating, in a single model, the COBE measurements and the comprehensive galaxy survey that yielded the large-scale structure data. The latest solution, hit upon by several groups, is to keep cold dark matter but to have some hot dark matter—perhaps massive neutrinos reincarnated—into the bargain. A mixture of 70 percent cold dark matter with 30 percent hot dark matter, in a universe that overall has the critical density demanded by inflation seems to do the trick—for the moment.

This, in short summary, is where things now stand. Inflation, and with it the stipulation that the universe must have exactly the critical density, remains the most popular general theory for the earliest moments of the big bang, even though the pieces of particle physics that have to be assembled to make inflation happen correctly are prodigiously complicated. A mixture of two types of dark matter, hot and cold, now seems to be the simplest way of explaining the galaxy distribution in its entirety, and even then a certain amount of biasing has to be thrown in to retain agreement with the smallness of the measured temperature fluctuations in the microwave background. Inflation, cold dark matter, hot dark matter, biasing—the terms are thrown about so casually by the builders of galaxy-formation theories that it is easy to forget that all of them are pure hypothesis, nothing but speculation. No one knows whether or not inflation actually happened a long time ago—or, indeed, whether or not particle physics happens to have the right form to make it fly. There is no such thing, strictly speaking, as cold dark matter or hot dark matter. Astronomers believe that dark matter of some form exists, but the particular types are theoretical inventions. Dark matter should be present right here on Earth, not just in distant space; it ought to be possible, with a suitable detector, to record fragments or particles of dark matter passing through any physics laboratory on the planet. However, despite a few recent attempts in that direction, no indication of it has yet been found. Biasing, the idea that dark matter and normal matter will have different dynamical behavior, is reasonable enough, but without knowing what kind of dark matter is up there around galaxies and within clusters, no one can say much more than that.

Modern cosmological theories are built on ideas that have no proven validity, if one insists on the old-fashioned standard of empirical evidence. The hope of the cosmologists is that, in the fullness of time, observations and theory will come together in one particularly neat arrangement so elegant that it will be persuasive despite the lack of solid evidence. Nothing in cosmology today encourages this hope. Simple theories have been replaced over the years by more complicated theories, which nevertheless do not work very well. Perhaps the universe really is a hugely complicated place; perhaps there are fifty-seven varieties of dark matter in the universe, coming from fifty-seven broken symmetries in particle physics. Perhaps there will never be a compellingly simple theory of galaxy formation, and we will just have to make do with the simplest theory we can think up.

In the meantime, cosmology and particle physics, having joined together because the practitioners of each discipline saw in the other new ways of asking and answering fundamental questions, have become inseparable. The relation between them is no longer one of mutual assistance but of absolute dependence. Cosmologists think cold dark matter is a safe bet because particle physicists can provide it; particle physicists invent theories containing cold dark matter because cosmologists seem to want it. Cosmologists think the universe must have exactly the critical density, despite the observational evidence, because inflation demands it; particle physicists think inflation is a marvelous idea because it predicts a universe of precisely critical density, which is what the cosmologists want. The hopes of cosmologists and particle physicists have become the same hopes, and their methods have become the same method: on both sides of this joint effort there is absolute reliance on the notion that a single theory will explain everything, and that when such a theory comes along it will be instantly recognizable by all. This might be called the messianic movement in fundamental science.

III

THE WHOLE WORLD

There is, we are aware, a philosophy that denies the infinite. There is also a philosophy, classified as pathologic, that denies the Sun; this philosophy is called blindness. To set up a theory that lacks a source of truth is an excellent example of blind assurance.

—Victor Hugo
Les Misérables, Book VII, ch. VI.

MANY THINGS in particle physics and cosmology that were once mysterious are now understood. Physicists believe that they know how the strong, weak, and electromagnetic forces can be unified; cosmologists believe that they have an idea how galaxies formed and were spread across the sky as they are. Together, they think they can begin to understand how the universe itself, and the particles in it, came to be as they are. They are pushing toward a theory of everything.

But the clues they need to guide them on this search are scarce. The microworld of particle physics has been exposed by particle accelerators, but those probes can go no further. The macroworld of cosmology will, no doubt, be explored further as bigger telescopes are built, but no amount of technical ingenuity will permit the cosmologists to see other universes, or our own universe at any time other than the present. What the searchers for a final theory are trying to find is something that can be truly grasped by the mind alone. And therein lies the problem.

8

March of the Superlatives

STEPHEN HAWKING'S celebrated lecture "Is the End of Theoretical Physics in Sight?"[1] was delivered on his inauguration in 1979 to the Lucasian chair of Mathematics at Cambridge University, a position most notably associated with Isaac Newton, and in more recent memory held by Paul Dirac. Newton was the instigator of the modern mathematical way of doing physics, and Dirac had endowed quantum mechanics with a formal, axiomatic style, elegantly stated, that aimed to deliver physical results—in Dirac's case, the prediction that antimatter should exist—from fundamental principles. Hawking, in his turn, came to the Lucasian chair at the age of thirty-seven having established his reputation as a mathematical physicist by constructing a number of powerful and far-reaching proofs in classical general relativity. With the Oxford mathematician Roger Penrose, he had proved that the beginning of our universe was a "singularity," a mathematical point of infinite density. Intimations of this initial singularity had existed, one way or another, since the 1920s, when general relativity was first used to build models of the expanding universe, but it was never clear whether the singularity was inevitable or a consequence of the simplifying assumptions that went into the models. It had sometimes been hoped that more complex mathematical universes would not contain the singularity, but Hawking and Penrose proved, using only the properties of general relativity along with some broad and easily proved statements about the universe today, that the singularity could not be escaped.

The only let-out from the singularity theorem of Hawking and

Penrose was that it was based on general relativity alone, and took no account of quantum mechanics. The uncertainty principle must forbid, it would seem, a true singularity. In quantum mechanics, no physical object can be given a precise location; it can only have a greater or lesser probability of being in one place or another. The singularity demanded by classical general relativity, on the other hand, was supposed to be an infinite amount of energy at a single point. The contest here between general relativity and quantum mechanics was a direct contradiction, like the proverbial irresistible force running up against an immovable object. When Hawking delivered his lecture on the end of theoretical physics, there was reason to think that grand unification, the joining together of the strong, weak, and electromagnetic interactions, was as good as accomplished, and that before too long gravity might be brought into the fold as well. With all the forces unified, any contradictions between the previously separate theories would necessarily be resolved and—voilá!—there would be a single, consistent account that could tell us what the beginning of the universe must have looked like. And there would be nothing left for theoretical physicists to do.

The lack of a theory joining gravity to the other forces was, and remains, the final obstacle to the continuing efforts of cosmologists and physicists to understand the birth of our universe, the event of the big bang. Knowing nuclear physics, they had been able to understand the universe at ten billion degrees and an age of one minute; with electroweak unification and then grand unification, they had been able to sweep much farther back in history, to a cosmic age of one trillion trillion trillionth of a second— the time when grand unification was, so they hoped, making matter and getting rid of antimatter, and when inflation might have made the universe big and smooth. But farther back than this they could not go, as long as general relativity and the other forces remained unwed. The best they could do was to determine the point at which their theories would no longer work—a moment when the size of the universe, as prescribed by general relativity, was as small as it could be, according to the uncertainty principle, for the amount of matter it contained. To go back farther would have general relativity implying that the universe was smaller than quantum mechanics would permit, a contradiction that made physics meaningless.

This moment is called the Planck time, because the numbers that define it come from simple combinations of the speed of light, the strength of gravity, and Planck's constant. The Planck time is one millionth of a trillionth of a trillionth of a trillionth of a second—a meaningless number, one might think, except that physicists believe that they know what it means. Loosely speaking, our ability to understand the physics of the universe ceases when it is younger than the Planck time. This statement actually is meaningless, because if we are admitting that we do not know what preceded the Planck time, we cannot say that it is a millionth of a trillionth of a trillionth of a trillionth of a second after anything in particular. A better formulation is to say that the universe emerges somehow from a Planck era, in which the controlling physical theory is unknown to us, and that when it emerges it resembles a universe one Planck time after a singularity, had there been one.

When Hawking and Penrose proved, within the confines of general relativity, that the singularity was not to be escaped, they also proved that the beginning of the universe would never be understood until a way was found to combine gravity with the other forces in a single consistent "theory of everything." To Hawking, and to many others, this was the last fundamental problem in theoretical physics. If a solution were found, there would be no more fundamental problems awaiting solution, and physics would be complete. Only computation would remain.

In his 1979 lecture, Hawking did not claim that a theory of everything existed, but he had in mind an idea that seemed to have the potential of becoming a theory of everything. The favored idea was called supergravity, and it was an elaboration of supersymmetry, the idea that fermions (particles with half-integral quantities of quantum-mechanical spin) and bosons (particles with integral spins) are not completely different kinds of creatures but similar objects in different guises.

If supersymmetry is of any relevance in the real world, it must be broken, because we do not see pairs of fermions and bosons, with equal masses, arrayed before us. In the original conception of these theories, supersymmetry was broken "globally"—that is, it happened everywhere at once, as if a giant cosmic switch had been thrown that abruptly and ubiquitously turned off the equivalence between fermions and bosons. This kind of global symmetry breaking is mathematically possible but is thought ugly by most physicists, who dislike the idea of some sort of physical change happening at the same time everywhere. Global symmetry breaking does not strictly contradict Einstein's prohibition of physical influences that travel faster than the speed of light, because it is "pre-wired" into the theory. But it strikes physicists as a violation of the spirit of relativity, and they much prefer theories in which symmetry breaking is done "locally"—that is, there is at every point some quantity that decides whether supersymmetry is on or off, and the breaking of supersymmetry can begin at one place and spread out, like ripples on a pond.

Local supersymmetry breaking is aesthetically preferable, but it leads to a further proliferation of particles. The quantity that measures whether supersymmetry is on or off can vary from place to place, and because it is a continuously variable quantity it can sustain waves that in quantum-mechanical parlance are equivalent to particles. The new ingredient that allows local supersymmetry breaking is itself a new dynamical medium, a new creature in the particle-physics world. In this case, the new particle seemed able to fill a slot that was open; it could be put to immediate use.

Attempts to think about what would be needed for a quantum theory of gravity—a theory reconciling general relativity with quantum mechanics—usually begin with a particle called the graviton. Just as the photon in the quantum version of Maxwell's electromagnetic theory is the particle equivalent of electromagnetic waves, so the graviton was supposed to be the particle equivalent of gravitational waves, best thought of as traveling ripples in the curvature of spacetime. This graviton was entirely hypothetical, but if gravity was to be incorporated into a truly unified theory that even remotely resembled electroweak theory and grand unification, it had to exist. The nature of gravitational waves requires the graviton to have two units of quantum spin, and it is therefore a boson. If supersymmetry, broken or not, is a part of the true theory of everything, then this boson must have a fermion cousin, and this doubly hypothetical object is called the gravitino, just as the superpartner to the photon is called the photino. A graviton, along with one or more types of gravitino, is precisely what popped out of the mathematical machinery when supersymmetry breaking was done locally. Therefore, it was concluded, locally broken supersymmetry must have something to do with gravity, and the theory was called supergravity—gravity, in other words, given a quantum-mechanical treatment and made supersymmetric.

But there was much more to the appeal of supergravity theories than the emergence of particles that looked like the graviton and the gravitino. The bugbear of all attempts to make quantum theories of gravity was that when anyone tried to calculate a force or a reaction rate using methods familiar from the way electromagnetism had been made into a quantum theory, the answers always came out infinite. The appearance of infinities was not as alarming as it might sound, because similar infinities had invariably poked their heads up in the first attempts to deal with electromagnetism by means of quantum mechanics. The problem, simply stated, is that in a quantum-mechanical calculation of the force between two electrons, for example, it is not enough to think of one electron at one place and the other at another, and then use the inverse-square law to work out the force between them. Quantum mechanically, each electron is described by a wavefunction that spreads throughout space, and the quantum calculation of the force between these two electrons involves figuring out the probability of one electron being here, one being there, working out the force for that particular placement, then adding up all the little bits of force that arise from all possible placements. This sounds complicated, but in practice it is not too bad, except for one thing: among the possible placements are instances where the two electrons are right on top of each other, in which case the force between them becomes infinite. This does not violate Pauli's exclusion principle, because the two electrons are not physically in the same place; however, their wavefunctions spread through the same volume of space, and therefore have some overlap. In any such calculation, the overlap of the wavefunctions inevitably gives rise to an infinite contribution to the total force, and for years, in the 1930s and 1940s, the development of a quantum theory of electromagnetism was stymied, because no one knew how to deal with the infinities.

A particularly galling aspect of these insidious infinities was that they upset attempts even to define what the mass and charge of an electron meant. To measure mass or charge meant, in practice, measuring the force an electron exerted on some other particle or on some piece of laboratory equipment. But this force always came out to be infinite, which seemed to imply that the measured mass or charge of an electron ought to be infinite. Clearly this was not the case, since classical physicists had been able to measure both quantities with precision, but infinite charge and infinite mass is what the quantum theory of electromagnetism invariably predicted.

In the seeming paradox, however, lay the solution to the technical difficulty. The calculations that produced infinite answers began with some "true" mass and charge for the electron, and then proceeded to work out how these numbers would be altered, by quantum effects, to yield an "apparent" mass and charge. These "apparent" values would be what the experimental physicists would measure—but they came out to be infinite. However, the basic philosophy of quantum mechanics is that the only numbers to be trusted are the ones that can be measured directly. The "apparent" values are, in this case, the only meaningful values in town, and the would-be "real" mass and charge of the electron, which went into the beginning of the calculation, are in fact meaningless, because they can never be measured. This being so, it seems allowable to pretend that what had been the "real" mass and charge of the electron are in fact infinite and negative; to be precise, they are whatever they need to be to cancel out the annoying infinite and positive quantities that crop up in the theoretical calculation. This gets rid of the infinities, and leaves an "apparent" mass and charge, both of which are well behaved and can be safely considered equal to the measured values.

What made this trick work was the serendipitous fact that once these two infinities, for mass and charge, had been disposed of, all the infinities that might arise in other calculations also vanished, because they turned out to be related in simple ways to the fundamental ones. With the process of "renormalization" accomplished, the offending infinities could be struck boldly from any quantum-mechanical calculation of an electromagnetic force. Now every question that could be asked had a sensible answer, and a new theory—quantum electrodynamics—was born.

Renormalization has been regarded differently by different physicists. Whether it is acceptable to say that the "real" mass and charge of the electron are infinite is, to the quantum-mechanical purist, a silly question, because the "real" quantities are ones that can never be seen or measured. They can be anything or nothing. To Paul Dirac, on the other hand, mathematical formalist that he was, the striking out of infinities was a distastefully pragmatic step, and he reckoned renormalization to be the great flaw of modern quantum theory. For most physicists, as always, pragmatism reigns. Quantum electrodynamics, suitably renormalized, works well and accurately; it can calculate experimentally measurable quantities to ten or more decimal places, once the infinities are deleted.

It would have been a happy outcome if this same procedure, so arduously acquired by physics in the middle of the century, could have been adapted to a quantum treatment of gravity. But this was, it was soon realized, impossible. Gravity is more complicated than electromagnetism. When two bodies are pulled apart against their gravitational attraction, energy must be expended, and if they come together energy is released; but energy, as Einstein so famously proved, is equivalent to mass, and mass is subject to gravity. Therefore, the energy involved in a gravitational interaction between bodies is itself subject to gravity. Gravity, if you like, gravitates. This makes calculations of the quantum-gravitational force between two bodies difficult, because when the force is calculated at one level of approximation, it feeds back into the calculation when the next level of approximation is attempted. The result of this is that many more infinities arise in this calculation than in the equivalent electromagnetic one, and the infinities are of a more recalcitrant kind. Whereas in quantum electrodynamics one subtraction of infinities suffices to make all the infinities disappear, in quantum gravidynamics (if there were such a thing) new infinities arise at every stage. Quantum gravity is non-renormalizable.

Supergravity, on the other hand, looked altogether renormalizable. Just as the photon, the quantum-mechanical particle corresponding to electromagnetic waves, is the basic unit for any description of any electromagnetic interaction, so the graviton is the medium of exchange for a quantum theory of gravity. A gravitational interaction between massive objects is described by the changing space curvature that is generated by the bodies but also acts upon them; the wiggling around of the curvature of space-time is equivalent to the motions of some set of gravitons, and any attempted quantum calculation of the force produced by those gravitons is beset by an infinity of infinities, and is quite unmanageable. In supergravity, however, there are not just gravitons but gravitinos, and any quantum theory includes the effects of both. The particular virtue of supergravity, so it seemed a few years ago, is that the infinities introduced by the gravitinos cancel out the infinities due to the graviton, leaving a renormalizable quantum theory of gravity. To be precise, the quantum theory of supergravity was better than renormalizable: the countervailing actions of the gravitons and the gravitinos suggested that no unpleasant infinities of any sort would arise in the first place, so there would never be any need to resort to the somewhat underhanded subtraction that was needed in quantum electrodynamics.

Better yet was the perception that supergravity led to an almost unique choice for the unified theory encompassing gravity and the other interactions. Supergravity is not one theory but a family of them, characterized by the number of distinct species of gravitino they contain. There was $N = 1$ supergravity, which contained but a single gravitino type, up to $N = 8$, which contained eight. (There are theories with more than eight gravitinos, as it happens, but they also contain new kinds of particles that are not wanted at all.) $N = 8$ supergravity contains not just the graviton and eight gravitinos (which have quantum spins of 2 and ½ respectively) but a stack of other particles: twenty-eight with one unit of spin, fifty-six with half a unit of spin, and seventy with zero spin.[2] From these particles, and these alone, the rest of the world— quarks, electrons, neutrinos, gluons—must be made. All the particles must be used for something, and no extras can be added, or the infinity-free virtues of the theory will be lost. The $N = 8$ version of supergravity was the particular theory that Hawking touted in his Lucasian lecture, and it is the only one with enough particles in it to be able, even in principle, to accommodate the observed particle content of the world we inhabit. It became, for this reason, a candidate of unique suitability to play the role of a theory of everything.

We must pause to digest one fact. The natural habitat of $N = 8$ supergravity is a universe of eleven dimensions—one dimension of time, and ten of space. These extra dimensions are there, in effect, to accommodate the supersymmetry of the theory, and a genuine geometrical significance can be attached to them. If we think of the basic unit of this theory as a little arrow in the ten dimensions of space, then we can say that when the arrow points along, say, the seventh dimension it represents a tau neutrino, but when it points along the eighth it is seen as a charmed quark; twist it into the ninth dimension, and it becomes a gluon.

The practical virtue of the theory is that its ability to dispose of all kinds of infinities comes about naturally, because every particle is related to every other particle and nothing is thrown in by hand. It is one thing to construct a quantum-gravitational theory by repeatedly adding new particles in order to dispose of the infinities created by the particles already there, and then adding more particles when further infinities arise at a more complex level of calculation. It is quite another to find, all in one piece, a theory that by its very nature contains a lot of intimately linked particles and therefore has no infinities. But it was not all plain sailing. First there was this eleven-dimensional world, which had to be accepted because it was the only world big enough to house $N = 8$ supergravity. In our world, only four dimensions (including time) are evident to our senses, but theoretical physicists had long ago got over the restrictive notion that what we see in the world must be taken at face value. As with all the broken symmetries that reconcile the differences between the strong, weak, and electromagnetic forces, and as with the broken supersymmetry that joins together seemingly unrelated fermions and bosons, it was not too hard for physicists to come up with a way in which the world could "really" be eleven-dimensional, even though we are privileged to see only four of them.

The idea of extra dimensions has a history, in fact, going back to early attempts to unite gravity and electromagnetism. In an early "theory of everything" proposed in the 1920s by the Polish physicist Theodor Kaluza, and later augmented by the Swede Oskar Klein, there was at every point in space a little circle, in addition to the normal three dimensions of space and one dimension of time. This was quite literally an extra dimension of space-time: to specify completely the position of a point in these five dimensions, one had to describe not only its position in the usual three dimensions, along with the time at which the measurement was being taken, but also the position of the point on the circumference of the little circle. Obviously, we do not perceive this little circle directly, which means it must be very small—in the 1920s, anything smaller than an atomic nucleus was satisfactory. What Kaluza and Klein tried to do was to make the position of the point on the circle represent the value of the electromagnetic field. In this way, electromagnetism was meant to be the geometry of the fifth dimension, just as gravity, according to Einstein, is the geometry of the four familiar dimensions. The Kaluza-Klein theory possessed a certain mathematical elegance, but it never really worked: it predicted that any object possessing, like the electron, a single unit of electric charge would have a far bigger mass than the real electron was known to have.

Still, the Kaluza-Klein theory had a certain persistent charm, and despite its total lack of correspondence with the empirical world, it lurked in the back of physicists' mind as an example of the sort of theory God (Einstein's God, anyway) might have used to put the world together, were God a mathematical physicist. When supergravity, with its insistence on an eleven-dimensional world, came on the scene, some physicists could easily believe that God had decided after all to make use of the elegant invention of Kaluza and Klein. Necessarily, the extra dimensions had to be made small somehow, so that they would not be obvious to us, but now that seven dimensions instead of one had to be hidden there were lots of ways to perform the concealment. The simplest was to make each of the seven extra dimensions a little circle, so that the extra dimensions became, in sum, a little seven-dimensional sphere. If this is the case, then each point of our conventional four-dimensional spacetime has to be regarded no longer as a point but as a little seven-sphere (as it is called), requiring an extra seven numbers (the equivalent of a set of latitudes and longitudes on a seven-dimensional globe) to say exactly where each point is. These seven numbers then say, in effect, what kind of particle is present: one combination of latitudes and longitudes is the specification of a muon, another combination specifies a quark, and a certain geometrical transformation—a particular rotation of the seven-sphere—turns the muon into the quark. This is the way in which supersymmetry makes all particles cousins by means of higher geometry.

One immediate worry, though, is that the seven-sphere is by no means the only possible arrangement of the extra dimensions. The radius of some of the extra dimensions can be made different, so that the seven-sphere turns into a kind of seven-dimensional football. Or the circles can be interlinked to make a seven dimensional doughnut instead of a seven-dimensional sphere or football.

While the general idea of an eleven-dimensional world did not altogether dismay the physicists, the fact that there was a large choice in the geometrical disposition of these extra dimensions meant that supergravity was not unique. Nothing in supergravity itself suggested that one way of rolling up the extra dimensions would be better than another, so the choice had to be made by physicists and not by nature. If no one way of rolling up the extra dimensions was fundamentally preferable to any other, the claim that supergravity could be a unique "theory of everything" had to be put aside. Either the physicists have to make the dimensions roll up in whatever way they find most convenient, or else they must live in hope of finding some higher theory, beyond supergravity, that dictates how supergravity is turned into an acceptable account of our particular universe, with its four dimensions and all its broken symmetries. Supergravity could not be a theory of everything after all. To Einstein's query as to whether or not God had any choice in the creation of the world the answer, evidently, was still, Yes, quite a lot.

In the end, there were also technical reasons for supergravity's failure to live up to Hawking's billing of it as the end of theoretical physics. The surpassing virtue of the theory was its apparent ability to do away with the infinities plaguing all attempts at realistic calculation in quantum-mechanical versions of gravity, but in fact this freedom from infinities had never been incontrovertibly established. What had happened originally was that while calculations involving a single graviton had thrown up the usual intractable sort of infinity, the addition of the gravitino produced, by virtue of the intrinsic symmetries of supergravity, equivalent infinities with the opposite sign: everything canceled out, and the theory produced sensible answers entirely of its own accord, with no need for the subterfuge of renormalization. But this was demonstrably true only for the one-graviton calculation, and there was no proof that the same sort of cancellation would so obligingly occur in calculations involving several gravitons and gravitinos. Supergravity contains so many particles and so many interconnections that performing calculations of interactions involving more than one graviton or gravitino required hours of effort and pages of scientific journals. Some examples were found of two-graviton calculations that appeared to be finite, but as the calculations became more complicated, infinities began to blossom in number and variety, like weeds in a well-kept garden. It looked increasingly less likely that the theory would actually work as intended, producing sensible answers free of infinities to all levels of calculation. In the end, supergravity died, choked by its weeds.

It is hard to find an obituary for supergravity in any scientific journal. Many physicist-hours were spent on supergravity, in efforts either to prove by direct mathematical demonstration that it was free of infinities at all levels, or to find examples of infinities that did not disappear quietly. The latter was the eventual outcome. No doubt some physicists paused briefly to reflect on the passing of what had seemed a wonderful idea, but no one took the time to eulogize. Scientists are not inclined to pause and reflect at the demise of a once-bright idea; if they experience any grief, they find release from it in action, not in mourning. Their idea of recovery is to move as quickly as possible to the next bright idea. And, as it happens, a new idea—another superidea— was waiting for them.

The new object of their affections was the superstring. Like super- gravity, this new object lived in a world of many more dimensions than four, but the rise and fall of supergravity made this circum- stance seem commonplace. The new theories did away with another well-established piece of common sense, however: they proclaimed that particles are not particles at all, but little strings, whose ends are tied together to form loops. Mathematically, this is an easy progression. Particles, idealized as points, are the sim- plest kind of object one can imagine, and strings, idealized as lines, are the next simplest. Since, with the demise of supergrav- ity, it was beginning to seem that all theories based on point parti- cles were doomed to failure because of the plague of infinities that afflicted them, strings were the next thing to turn to. In string theory, all infinities of the traditional kind are banished at the source. If particles are in fact little strings, their quantum- mechanical attributes—mass, charge, strangeness, charm, and so on—are spread out along the string instead of being concentrated wholly at one point, and do not produce an infinite concentration anywhere. In any quantum-mechanical calculation of the behav- ior of two strings, the respective wavefunctions still overlap, but the spreading out of charge and mass and everything else insures that the overlapping creates no infinities. String theory is, by its very nature, free of infinities.

Modern string theory has a curious and, by the standards of particle physics, venerable pedigree. Back in 1968, in an attempt to understand the then-puzzling question of why quarks could not escape from the confines of neutrons and protons, the Italian physicist Gabriele Veneziano came up with a string theory: quarks were not particles but the ends of pieces of string, and, naturally, if you cut a piece of string in two you create more quarks. Quarks could never be seen in isolation because they are not particles at all. A piece of string cannot have just one end. Veneziano's string theory never worked except in a cartoonlike sort of way; one obvious problem was that quarks have to be bound together in threes within neutrons and protons, which would require a string with three ends. In the 1970s, modern quark theory took hold, and the confining effects of gluons, which provide a force that increases as quarks are pulled apart, explained the nonexistence of single quarks. Veneziano's theory fell by the wayside. But ideas in theoretical physics, once brought into the world, can be tenacious, and two physicists—John Schwarz, of the California Institute of Technology, and Michael Green, of Queen Mary College in London—kept string theory alive, at least as an exercise in formal mathematical theory.

A curious property of strings, recognized early, was that they seemed much happier living in a large number of dimensions. Just as in quantum terms electromagnetic waves are translated into photons and gravitational waves are reconstructed as gravitons, so the basic quantum-mechanical units in string theory are waves that run along the string itself, like the vibrations of a violin string. These are the particles of string theory, and they must ultimately be identified with the particles we observe in our accelerators. But these string vibrations can have some curious properties. A string moving through four-dimensional spacetime sweeps out a two-dimensional surface, a rippled sheet. The ripples are the fundamental string vibrations that, in the quantum theory, become the elementary particles of the string theory. Alarmingly, however, some of these ripples translate into particles that move faster than the speed of light.

Faster-than-light particles are popular in science fiction, and have been toyed with over the years by a few physicists, but they are basically a nuisance, giving rise to physical processes that make no sense. In quantum mechanics, faster-than-light particles generate yet another kind of infinity, in which events can influence themselves by propagating some effect that travels back in time and then forward again to alter the very process that generated it. (This is the equivalent, for elementary particles, of the son who goes back in time to kill his father). String theories suffer badly from faster-than-light effects, except, it turns out, in the special case of strings that live in twenty-six dimensions (twenty-five space dimensions, plus time). Uniquely, twenty-six-dimensional string theory has no particles that can kill their parents. There is no simple explanation of why this should happen; it is a free gift, provided by the gods of mathematics. A twenty-six-dimensional world is an odd thing to be thinking about, and most people, if they thought about it at all, were inclined to regard the benignity of string theory in twenty-six dimensions as one of the many odd but useless facts that bring joy to the hearts of mathematicians and puzzlement to everyone else.

A second curious fact of string theory attracted attention. It had been noticed, back in the days of the Veneziano model for quark confinement, that among the fundamental string oscillations there was always one with two units of quantum-mechanical spin. This was unwanted in any quark theory, but John Schwarz and Michael Green, among others, saw the hint of a connection with gravity: the quantum-mechanical messenger of the gravitational force, the graviton, also has two units of spin. This was the first inkling that string theory might have wider import than had first been appreciated.

The pure twenty-six-dimensional string theory suffered from a considerable problem, however, in that all the string vibrations—the quantum-mechanical objects of the theory—were particles whose spin came in whole-number units: the theory had lots of bosons and not a single fermion. This gave it some faint credibility as the basis for a unified theory of interactions, since all the force-carrying particles—the photon and graviton, the gluons, the W and Z particles, and the X and Y—are bosons. But a theory containing nothing but bosons seemed like a poor shot at a theory of everything.

Then came two events, both in 1984, that transformed string theory from a plaything of excessively mathematical physicists into the heartthrob of every particle physicist and would-be unifier of gravity and the other interactions. First, David Gross, Jeffrey Harvey, Emil Martinec, and Ryan Rohm (later dubbed the Princeton string quartet) invented a derivative of the twenty-six-dimensional string which deigned to live in a mere ten dimensions, the other sixteen having been, in effect, wrapped up in the structure of the string itself. A simple string in twenty-six dimensions was transformed into a more complicated string in ten dimensions, but the bonus of this transformation was that the internal structure of the string (the offspring of the sixteen former dimensions) became the geometry of supersymmetry. Thus the string became the superstring, and because supersymmetry ruled its physics, every oscillation that manifested itself as a boson had to acquire a partner in the form of a fermion. If all was done correctly, in fact, supergravity, the recently deceased theory of everything, was now seen as merely a superficial attribute of the deeper physics of the superstring.

The second event of 1984 was a more subtle realization, due to Schwarz and Green themselves. It involved an odd fact of the physical world, which had first turned up in the 1950s. Until that time, it had been taken for granted that one of the inevitable symmetries of the physical world was mirror symmetry—that is, any physical process is indistinguishable from (proceeds at the same rate as, obeys the same laws as) its mirror image. This seemed almost a self-evident proposition, but in 1956 the Chinese-American physicist C. S. Wu conducted a difficult and historic experiment showing that as far as beta decay was concerned mirror symmetry was not a law of physics. Wu took radioactive cobalt-60 atoms, which possess one unit of spin, and lined them up in a magnetic field so that their spin axes were all in the same direction—to be specific, imagine that the cobalt atoms are like right-handed screws, placed point up. When a cobalt atom decays, it produces an electron and an antineutrino, and Wu recorded the directions in which the electrons were emitted relative to the alignment of the cobalt atoms. She found that the electrons were emitted predominantly downward—that is, opposite to the spin direction of the cobalt atoms. This doesn't sound surprising, but it is. Imagine a mirror image of these cobalt atoms: the electrons still predominantly go downward, but the mirror has turned the right-handed spins of the cobalt atoms into left-handed spins, which are represented by arrows pointing down rather than up. In the mirror-image version of this experiment, therefore, the electrons come out preferentially aligned with the direction of the atomic spin and not opposite to it. Beta decay, evidently, could tell left-handedness from right-handedness in particles, which implied that the weak interaction, the driving force behind beta decay, distinguishes particles from their mirror images.

The unnerving handedness of the weak interaction is not an easy thing to incorporate into physical theories. There are many mathematically allowed theories of particle interactions which cannot accommodate handedness at all; any attempt to force it upon them causes all kinds of fatal disasters, including breakdowns of the law of conservation of energy.

The need to accommodate handedness in the weak interaction thus becomes an acid test of all putative theories of everything. The application of this test to superstring theory produced a remarkable result. The original twenty-six-dimensional string theory was unique: the string was a featureless line wiggling around in twenty-six-dimensional space. But the transformation of the twenty-six-dimensional string into the ten-dimensional string with supersymmetry is far from unique; the sixteen dimensions rolled up in the internal structure of the string can be rolled up in an enormous number of ways. This is a matter of straightforward geometry, just as it was for the rolling up of the extra dimensions of eleven-dimensional supergravity theories to get the required four-dimensional world. If the aim is to find a unique theory of everything, in the form of a special ten-dimensional string theory preferred to all others, this abundance of possibilities does not look promising.

But at this point came the second great advance of 1984. Of all the possible ten-dimensional string theories, Green and Schwarz were able to find only one that could accommodate handedness in the interactions without breaking down. This seemingly unique theory was given the cryptic designation SO(32), which arose from the mathematicians' enumeration scheme for all the geometrical ways of rolling up sixteen of the twenty-six dimensions of the original string to get a ten-dimensional superstring. For the happy result that out of all the mathematically possible superstring theories only one had the capacity for handedness, physicists could only thank, once again, the gods of mathematics. It was a striking fact, all the more striking for being quite inexplicable. There was no reason why a theory of formerly twenty-six-dimensional strings wrapped into ten dimensions and incorporating supersymmetry in a certain specific way should turn out to be the only theory possible. The whole thing was a gift to physics from the world of pure mathematics, an answer to Einstein's desire to know whether "God had any choice in the creation of the world." The answer, it seemed, was, No: God had to use the SO(32) string in ten dimensions, just like the rest of us.

This moment of perfect fulfillment was sadly brief. A little later, a second string theory, the equally cryptic $E_8 \times E_8$, was also found to be able to accommodate handedness, thereby passing the same acid test. There was a story going around physics departments at the time that Richard Feynman had been quite enamored of string theory when he heard about Schwarz's and Green's result, but that when he heard that two such theories were possible he consigned the whole enterprise to the growing junk heap of would-be theories of everything. Despite Feynman's skepticism, however, string theories have not proliferated. For most physicists it remains an astonishing fact, impossible to ignore, that an elementary application of geometrical laws should select only two theories from the multitude of possibilities. When so remarkable a result comes with such unanticipated force into the minds of theoretical physicists, must we not accept that something out of the ordinary has been discovered? How can superstring theory not be the foundation for a theory of everything?

At this point, however, the difficulties of the "top-down" approach to doing physics can no longer be skirted. We should review briefly what has brought us to this juncture in theoretical physics. The demi-uniqueness of string theories arises from the application of general principles: any string theory is free of infinities of the sort that have to be "renormalized" away in quantum electrodynamics and the other point-particle quantum theories; of string theories, only the twenty-six-dimensional one is free of faster-than-light phenomena; this theory contains only bosons, but can be reduced to a ten-dimensional superstring theory that contains both fermions and bosons, related by supersymmetry; this reduction can be be done in an enormous number of ways, but of all these only the SO(32) and $E_8 \times E_8$ ten-dimensional strings successfully accommodate interactions with handedness. Finally, superstring theories contain bosons with two units of spin, which can be identified with gravitons, allowing the possibility of a connection to a quantum theory of gravity and thence to a classical theory of gravity resembling general relativity. These are all desirable and perhaps essential characteristics, qualities that any theory of everything must, at a minimum, possess. But where is the real physics in this? Where are the quarks and electrons; the strong, weak, and electromagnetic interactions; the slight asymmetry between matter and antimatter that fills our universe with protons but keeps antiprotons out? What is missing, so far, is a precise account of how the ten-dimensional superstring, the presumed basis of a theory of everything, reproduces the world we see around us.

The superstring theory is ten-dimensional but must, of course, be brought down to four dimensions before we can make use of it. Six dimensions therefore have to be rolled up into a small, compact, concealed geometry. We are getting used to this, and it comes as no surprise to learn that there are ways of doing it. But there are lots of ways of doing it. There is nothing in any part of string theory, so far as anyone knows, that makes a four-dimensional world special: the superstring itself is quite happy living in ten dimensions, and left to its own devices would stay there; it can be "compactified"—that is, six of its dimensions can be rolled up to yield an apparently four-dimensional world—but it could just as easily be compactified to make a three-dimensional world, or an eight-dimensional world. Superstring theory does not naturally make a four-dimensional world; physicists have to compel it to do so. And even after so compelling it, they find that there are tens of thousands of ways of rolling up six dimensions, with nothing to suggest one way over another. Here again, so far as we know, physicists have to do the choosing.

Next problem: superstring theory contains, in the form of string vibrations, lots of particles, which will eventually be identified with the quarks and the gluons, the neutrinos, the electron and muon, the photon and W and Z and Higgs bosons, and so on. But all the superstring particles are, in their pristine form, massless. We know how to give mass to massless particles, of course; that is what the Higgs mechanism is for. So mass can be provided where it is needed, and used to break the supersymmetry of the superstring so that fermions and bosons do not all come out in identical pairs. Then the Higgs mechanism must be used again, to break the symmetry between the strong and the electroweak interactions, and again to break the symmetry between the weak and electro-magnetic interactions. Finally, there will emerge the world we want, the world we happen to find ourselves in.

Physicists, in a general way, know how to do all this. Theoreti-cal progress in particle physics during the second half of this cen-tury has consisted in spotting partial or broken symmetries in the interactions studied at accelerators, creating theories that embody the symmetry perfectly, and then modifying the theory so that the perfect symmetry is concealed and only a remnant of it remains for us to see. This method has given us the quark theory for the internal structure of the baryons and mesons, the electroweak theory for the unification of the weak and electromagnetic inter-actions, and perhaps grand unification too. String theory is this idea taken to perfection: in the ten-dimensional world of the superstring, all particles are massless and all interactions are equal. But we do not live in that world. To get our mundane habi-tation, the perfection must be thoroughly chopped to pieces and buried in a mass of symmetry breakings. The ten dimensions must be reduced to four, the particles must be given masses, the inter-actions must be split apart and made different. Any uniqueness or perfection possessed by string theory exists, in truth, only in that prelapsarian ten-dimensional world. Everything that must be done to get from there to here, from ten-dimensional perfect symmetry to four-dimensional confusion, must be done by the physicists. The superstring theory itself is silent on that issue; it does not say how it should be broken up and disguised to meet our particular needs. Superstring theory may be a perfect, unique creation, a singular piece of mathematics. But it contains, appar-ently, the makings of a million different imperfect worlds, all equally possible. So far, physicists have no idea what specific ornamentation must be added to superstring theory to create the real world, let alone why that ornamentation rather than some other is preferred by nature.

Papers on superstring theories still fill the physics journals, but the sense of imminent triumph, the idea that the end of theoretical physics is in sight, has largely evaporated in the face of what might be called technical complications. But, as is often said, God is in the details—the technical complications. It still seems, in a general sense, that a string theory could be the basis for a theory of everything, but the idea that such a theory would explain (like the philosophy of Pythagoras) every last thing about the real world in terms of the laws and dictates of mathematics has faded.[3] As philosophers have frequently found, the real world seems too messy, too stubbornly arbitrary, to be found out by the power of thought alone, no matter how fine the guiding sense of aesthetics.

9

The New Heat Death

TOWARD the end of the nineteenth century, when the principles of thermodynamics and statistical mechanics had been securely established, and when the futility of trying to make perpetual-motion machines or steam engines that converted heat to mechanical effort with perfect efficiency had been understood and accepted, an alarming consequence for the universe as a whole was noticed. Rudolf Clausius, the German physicist who had invented the idea of entropy (and, along with it, the notion that all machines are unavoidably inefficient), observed that the whole universe should properly be regarded as a thermodynamic system of grand design. This had a distressing consequence. If the entire universe was a single, closed thermodynamic system (which, as a matter of definition, it had to be, because the universe contains everything that exists), then its total entropy must be increasing, and its store of available energy must be decreasing. The universe itself must be running down.

In our corner of the universe, the Solar System, most of the Sun's energy goes out into empty space, but a small fraction reaches the Earth. Solar energy striking our planet is, in the long run, responsible for everything that happens here. It heats the atmosphere and seas, creating air movements and ocean circulations that traverse the globe and control our climate. Some of the Sun's energy goes into plant growth, which cleans and recycles the atmosphere and sustains the lower reaches of the food chain, on which we ourselves depend. Hundreds of millions of years ago, plants grew and died in such profusion that their decayed remains, compressed by layers of rock, turned into thick seams of coal and huge subterranean oil reserves, which we now dig up and burn to make ourselves comfortable. We use hydroelectric power, wind power, and sometimes solar power directly to maintain our profligate civilization. All this ultimately derives from the Sun. (Nuclear power is almost an exception; the presence on Earth of uranium, from which we can extract energy, has nothing to do with our Sun, but it was manufactured by other suns long ago and mixed back into the interstellar gas when those earlier suns exploded as supernovae). Every source of motive power, therefore, comes from our Sun or from other suns, and we are using it up, transforming it into heat and electricity. And in the process, as Clausius proved more than a century ago, some of the energy is wasted. Our waste heat, for the most part, radiates away from the Earth and into space. The energy from the Sun which strikes the Earth is only temporarily waylaid, and inefficiently used, before it, too, leaks out into space.

Throughout the galaxy there are billions of stars, burning up their nuclear fuel to make heat and light, which race away into the darkness of interstellar space. Some of this energy leaks out of the galaxy altogether: the reason we can see other galaxies in the heavens is that some of their light travels to our eyes. Part of the energy is intercepted by the gas and dust that thinly pervade the space between the stars in our galaxy. The gas and dust are warmed by starlight, but only feebly. The stars will die when they use up their allotment of hydrogen fuel; some will explode as supernovae, scattering their contents into space as hot gas, which radiates its energy away and cools. From this gas some stars form, but these stars are made of partly used gas, containing less hydrogen than the pristine gas from which the first stars formed. This cycle cannot continue forever; fuel is used up, energy is generated and then dissipated.

When there is no fuel left from which more energy can be made, all physical activity will grind to a halt. Heat can move from one place to another, but overall, as dictated by the second law of thermodynamics, the net effect is always to equalize. Hot things get cooler and cool things get hotter, until everything in the galaxy—every dead star, every planet, every rock, and every atom of gas—is at the same temperature. At this point, no further activity is possible, because none of the energy is usable. Entropy has reached its maximum possible value. With every atom at the same temperature, there can be no net movement of energy from one place to another, and all physical processes must stop.

In Clausius' day, it was assumed that the universe was nothing more than a finite collection of stars, and he argued, from the laws of thermodynamics alone, that one day all possible sources of energy would have been used up, all temperature differences would have been erased, and all activity would cease. The universe would be, in the end, a warm, featureless, perfectly mixed bath of dust and gas. This dismal suffocation was called by Clausius the "heat death" of the universe. From the microscopic point of view, it can be understood as the inevitable equalization of all atomic motion. As cosmic history unfolds, every atom exchanges energy, directly or indirectly, with every other atom, and when all the atoms in the universe possess equal shares of energy macroscopic physical change is impossible: atoms will still collide, chemical reactions will still go on, but every reaction will be counterbalanced by an equal and opposite reaction, because overall the equilibrium must be maintained. There is motion still, and a perpetual interchange of energy among the teeming atoms, but these are only random, meaningless movements.

This vision of the death of the universe was distinctly upsetting to a generation of nineteenth-century thinkers and doers, who were getting accustomed to the idea that mastery of science bestowed on them a limitless capacity for self-improvement. But for nineteenth-century scientists the notion of the heat death raised a question about the past more perplexing than any about the future. If physics implied a constant degeneration of energy sources, an inevitable loss of the motive power that kept physical, geological, and biological change going, then it also implied a finite past. At the end of the nineteenth century, the universe was evidently a long way from the heat death, full of hot stars and cool planets strewn about in frigid space. There was a lot more thermodynamic activity to be enjoyed before complete uniformity and equality set in. The universe, by thermodynamic standards, was young. It had a relatively brief past; did it, therefore have a beginning? And should this beginning be the province of physicists rather than theologians?

Since the 1920s and the work of Georges Lemaître and Alexander Friedmann, physicists have pushed back to the point where now the only barrier to the comprehension of the very beginning of the universe is the inconsistency between general relativity and quantum mechanics. Now that the universe is seen as expanding and not static, the classical picture of the heat death has had to be revised. If the universe is open, and expands forever, thermodynamic activity will still, as Clausius imagined, grind to a halt, but all the energy created and radiated away by stars in their lifetimes will be redshifted, reduced, diluted by the expansion of the universe. The new "heat death" will be cold; no new energy will be created, and what is already there will be attenuated without limit. If, on the other hand, the universe is closed, and one day ceases its expansion and begins to contract, then in time everything will be squeezed together in a final "big crunch," a big bang with the movie projector running backward; conceivably, the cycle will start over again. Perhaps there will be an infinite future, if a new universe can be born from the remnants of the old. But whether the universe is open or closed, whether it expands forever or one day begins to fall in on itself, depends on the conditions of its birth, which we still do not understand. To know the fate of the universe we must understand its creation, and for that, again, we need to know how gravity is joined to quantum mechanics.

Superstring theory, even allowing that it has something to do with the unification of all the fundamental forces, has so far made not a single prediction about the structure of the universe, let alone about its origins. It is not at all clear how superstring theory could be made to say anything about the birth of the universe: it is the embodiment of the particle physicist's picture of the world, a theory built of discrete fundamental objects and the interactions between them. It pictures gravity as, in essence, the sum total of all the physical interactions and characteristics of a particle called the graviton. Where is general relativity in this? Where is geometry, curvature?

The graviton comes from the attempts to put gravitational waves—ripples in the curvature of spacetime—into quantum-mechanical language. In simple terms, any small region of spacetime can be regarded as being a flat section of spacetime, slightly warped or rippled; these ripples can be interpreted as gravitons, so that, in this quantum view, curved space is flat space plus gravitons. On the small scale this is fine, but when each patch of spacetime is pictured as a piece of flat space with some gravitons the larger geometry is lost to view. A patch of curved spacetime may belong to a closed universe, or it may be part of the space surrounding a black hole in an open universe. The graviton picture of quantum gravity is, in short, a local theory of particles and interactions; it is supposed to deal with objects and the forces between them at close range, and it is far from obvious how the larger geometry of the universe as a whole is tied into the small-scale picture.

There is a more subtle problem, too. Gravitons are supposed to move around in flat spacetime, and thereby create what we call gravity. But where did this flat spacetime come from? What ordained it? The whole point of general relativity is that the presence of matter creates the curvature of space, but in the particle-physics theories, superstrings included, the particles or strings are waggling around in a space that already exists—an absolute, almost Newtonian space that is there just because it's there. In this sense, all the quantum theories of gravity based on the properties and interactions of gravitons seem to be missing the very ingredient that makes general relativity work.

There may well be more to this than philosophy. Just as a naïve extrapolation of the expanding universe backward in time leads to a moment when the big bang, defined strictly by general relativity, is smaller than quantum mechanics and the uncertainty principle can possibly allow, so there is reason to think that in any marriage between general relativity and quantum mechanics the very notion of space and time breaks down at some submicroscopic level. It is not too hard to think of superstrings reeling around in spacetime, as long as the spacetime is still the classical, smooth, continuous empty space we have always imagined it to be. But if at some very small scale even length and time cease to have meaning, because the uncertainty principle does not allow infinitely fine subdivision, then how are we to think of spacetime at all, still less of the superstrings that are meant to inhabit it?

Superstring theory is, like most attempts at a theory of everything, an assault made using the strategy of particle physics: first reduce gravity to a theory of particle interactions—gravitons— then seek to incorporate this new particle into the elaborate machinery that particle physicists have built over the years. It is not obvious that this approach is ever going to work; certainly, it is hard to see how one is to imagine the birth of the universe in such terms. Because of these doubts, assaults on the problem of quantum gravity have been made from the other direction, by using general relativity and curved spacetime in all its geometrical complexity as the ideal and attempting to weld quantum mechanics onto it. This less popular approach has, so far, produced the only achievement that can almost be counted as a quantitative, testable piece of new physics. This is Stephen Hawking's demonstration that black holes are not quite black after all.

To the astrophysicist, a black hole is exactly what its name implies—a perfectly efficient vacuum cleaner that sucks material up and never lets it out. Black holes form when so much mass gathers in one place that no source of internal energy can hold it up against its own gravity; when the most massive stars run out of nuclear fuel, for example, their cores are thought to turn into black holes that may contain several times the mass of the Sun compressed into an object only a few miles across. The surface of such a black hole, called its "event horizon," is the boundary from behind which light can never escape the pull of the black hole's gravity. An object that falls through the event horizon is lost forever. But even though we cannot see what goes on inside this skin, theoretical physicists can think about it: everything falls to the very center, and accumulates there. The entire mass of the black hole seems to reside at a mathematical point, dead center, resembling the singularity with which Hawking and Roger Penrose proved that the universe began.

But this conclusion follows from general relativity alone, and, as before, it is to be supposed that quantum mechanics somehow prevents the formation of a true, mathematical singularity. In 1974, Hawking made an effort to understand what black holes would look like for a quantum-mechanical rather than a classical observer. To his and everyone else's surprise, he found that they were not perfectly black and absorbent, but leaked a little. The smaller the black hole, the more it would leak, and Hawking showed that black holes have a temperature, inversely proportional to their masses, that determines the rate at which they leak particles to the outside world.

This remarkable conclusion can be understood, retrospectively, in a simpleminded way. The event horizon of a black hole is supposed to be a perfect one-way barrier enclosing a volume of space: it lets things in but allows nothing out. According to quantum mechanics, however, perfect barriers cannot exist. Any quantum-mechanical particle should be imagined not as a point but as an extended wavefunction that measures the particle's probability of seeming to be at this place or that. Even when a quantum-mechanical particle is swallowed by a black hole, its wavefunction has some presence beyond the black hole's event horizon as well as within it. Any particle "inside" the black hole, therefore, has some finite probability of appearing outside it, because there is always a little tail-end of its wavefunction wagging feebly outside the event horizon. The smaller the black hole, the smaller its event horizon, and the greater the chance of particles being able to sneak out. Small black holes are not so black as big ones.

This loose argument could have been made, unconvincingly, in the 1960s, when the physics of black holes was first being explored. What Hawking did in 1974 was to make it mathematically secure, to turn a vague specter of an idea into a quantitative theory. In particular, he showed that the presence of the singularity at the black hole's center was essential for allowing the "virtual" particles, represented by the wavefunction beyond the event horizon, to turn into real particles that could move away from the black hole and be detected by familiar means.

In practical terms, the fact that black holes were not perfectly black was of negligible importance: the smallest conceivable black hole that can form today is one with something like the Sun's mass, and its temperature, according to Hawking's theory, would be a ten-millionth of a degree above absolute zero. Space is filled with the cosmic background radiation, at three degrees above absolute zero, so such a black hole would be cooler than its sur- roundings and would absorb heat from the universe faster than it radiated energy away—it would grow, not shrink. But in the world of theoretical physics, Hawking's discovery was a true reve- lation. It was the first time that a conjunction of general relativity and quantum mechanics had produced real, and therefore poten- tially observable, consequences. A black hole small enough to throw off particles at any detectable rate would have to be extremely tiny: it might weigh a billion tons, but that mass would be concealed within an event horizon the size of an atomic nucleus. It had been suggested that such an object might have formed in the very early moments of the universe, and some ten- tative searches were made for the characteristic gamma-rays and X-rays that such a tiny and very hot black hole would generate. No empirical evidence to support Hawking's idea of black holes that radiate has been found, but Hawking's argument was as real a result as one could hope for in this kind of physics.

It also came from an unexpected quarter. Hawking was renowned for his work in classical general relativity, whereas most of those engaged in the business of trying to reconcile quan- tum mechanics and gravity came from the world of elementary- particle physics, and regarded the problem of unifying gravity with the other forces as the search for a consistent, infinity-free theory of gravitons. Hawking's idea of black-hole radiation, by contrast, employed a sophisticated understanding of relativity and curved space but used only the bare principles of quantum mechanics. It was a genuine mix of general relativity and quan- tum theory, but it had nothing to do with the unification theories proposed by the particle physicists.

Hawking's technique in attacking the black-hole problem may, however, have further uses. If the singularity inside a black hole can be dissected in this way, what about the singularity at the beginning of the universe? Pursuit of this idea, mostly by Hawking and his collaborators, has produced a subject known as "quantum cosmology." In classical cosmology, gravity controls the macrocosmic expansion of the universe and particle physics controls the microcosmic processes within it, but otherwise the two have nothing to do with each other. This separation cannot be sustained at the very earliest moments, when the size of the universe is comparable to the range over which the particle interactions act. This is where the need for a unified theory of gravity and particle physics becomes inescapable.

Superstrings and other potential theories of everything are supposed to be capable of handling this problem, but they are a long way from it. In them, gravity is reduced to the collective action of gravitons, and the overall geometry of an expanding universe does not emerge from this sort of theory in any obvious way. Quantum cosmologists, therefore, take a different tack. Instead of worrying about how to combine all the particle interactions with gravity, and then trying to figure out what the combination implies for the beginning of the universe, Hawking and his followers throw away all the complications of the strong, weak, and electromagnetic interactions and suppose that a more direct way to understand the birth of the universe is to think about the implications of fundamental quantum-mechanical principles for gravity alone. They are attempting to do for the universe as a whole what Erwin Schrödinger did in the 1920s for the hydrogen atom. His analysis of the behavior of an electron orbiting a proton replaced the classical electrons, pictured as little planets following strictly defined orbits, with electron wavefunctions, whose intensity at any point denoted the chance of finding an electron there.

For classical electron orbits, now think instead of classical cosmological models. General relativity allows a variety of cosmologies: some expand from a point to a maximum size then back to a point again; others expand forever. More generally, if the strict assumption of cosmic uniformity is dropped, theoretical universes can expand faster in one direction than in another, or can contain denser spots, which fall in on themselves even while the rest of the universe continues to expand. There is an infinite variety, a cornucopia of nonuniform universes. The history of all these possible universes can be pictured as lines drawn in some higher dimension; a single point on one of these lines represents a particular cosmic geometry at a particular moment, and an evolving universe traces out a continuous but perhaps tortuous path. As Schrödinger replaced electron orbits with wavefunctions that described the probability of an electron doing one thing or another, so the quantum cosmologists want to replace the pencil line of an individual cosmological trajectory with a wavefunction that indicates the probability of the universe having one particular geometry or another.

Quantum cosmology ought to yield probabilities. Schrödinger's equation says that the electron is more likely to be found in this place than that, and the corresponding equation of the quantum cosmologist should say that one kind of universe is more probable than the others. The hope, of course, is that simple uniform universes, of the sort that theorists have been dealing with for years and in which we appear to live, will turn out to be the most probable.

This Panglossian approach sounds excessively simple, and of course the actual analysis that follows from this general idea is complicated and subject to numerous assumptions. One in particular stands out. Schrödinger's equation, applied to the hydrogen atom, does not automatically produce the complete description of the electron in orbit. Solutions to the equation are of two sorts: one kind describes an electron whose presence is concentrated around the nucleus and decreases in probability at greater and greater distances, but in the other sort of solution the electron's presence increases in probability the greater the distance from the nucleus. This second kind of solution is clearly not a description of a stable, isolated hydrogen atom, and to dispose of it the physicist applies what is called a "boundary condition." Any solution of Schrödinger's equation applicable to a real hydrogen atom must describe an electron whose presence diminishes with distance from the nucleus. The boundary condition, simply stated, is that the wavefunction must vanish at infinite distance; anything that has some finite magnitude an infinite distance from the atomic nucleus is disallowed—it is unrealistic. The well-behaved solutions that remain after the others have been discarded then give a correct description of the behavior of electrons in real hydrogen atoms.

It is a standard feature of most of the equations of physics that they require boundary conditions for the appropriate solution to be picked out, and there is nothing underhanded about using boundary conditions to make the right selection. The boundary condition is derived from the physical situation the equation describes, and it is just as much a part of the physics as the equation itself. Throwing away the bad solutions to Schrödinger's equation because they do not fit the physics being modeled is perfectly legitimate.

In quantum cosmology, inevitably, a similar need for boundary conditions arises, but the strategy for deciding which solutions to keep and which to throw away is far from clear. Hawking has got around the difficulty by promulgating an idea he calls the "no-boundary condition," which amounts to a declaration that the only universes that make sense are ones with no boundaries, either in space or time.[1] Closed universes satisfy this restriction. Like the surface of the Earth, they are finite but have no "edge"; because they expand from a singularity, come to a halt, and fall back to a singularity, they are finite in time, too. (Depicted this way, closed universes have a beginning and an end, and therefore have boundaries in time, but by a mathematical transformation the quantum cosmologists make the initial and final singularities of the closed universe into the same point in spacetime, so that, as T. S. Eliot had it, in their beginning is their end, and there is no boundary.) If the restriction to universes with no boundaries is made, then Hawking has demonstrated that uniform universes are by far the most probable. Quantum cosmology, then, appears to predict that the universe is both closed and uniform—which it might indeed be, although cosmological observations are far from deciding the matter one way or the other.

In thinking about the birth of the universe along with the quantum cosmologists, we must do away with a pointlike singularity of infinite density, as general relativity on its own implies, and imagine instead a universe passing from quantum to classical form, emerging from a literally uncertain initial state into the predictable, dependable universe of classical physics. The universe solidifies, so to speak, from an ectoplasm of quantum uncertainty, and once congealed proceeds to evolve in a deterministic, classical manner. Quantum cosmology supplies the method for picking out, from all possible pre-classical universes, the most probable, and if the no-boundary condition is applied, closed and uniform universes are what to expect.

To say, in Hawking's phrase, that the boundary condition for quantum cosmology is that there should be no boundary makes it sound as if the need for boundary conditions has been obviated. In truth, however, the decision to restrict attention to boundary-less universes is, in its own way, a boundary condition—a way of getting rid of unwanted solutions to the equations. For the hydrogen atom, we are justified in throwing away the bad wavefunctions, because we know in advance what hydrogen atoms actually look like, but for quantum cosmology there is no comparable independent check on the choice of boundary condition. If we were able to survey a great number of universes, as we can hydrogen atoms, we could figure out the correct conditions empirically. We have only one universe to examine, however, and the whole point of doing quantum cosmology was to find out from first principles what the universe ought to look like. This does not work: only by making an additional assumption, in the form of the no-boundary condition, does a sensible answer emerge, but the reason that the particular assumption was made was because we wanted to get a sensible answer. Clearly, this is circular reasoning. Hawking throws away "bad" universes because his quantum-cosmological formulation would go awry if they were included. But deciding that certain phenomena do not exist because the available theory cannot cope with them is a back-to-front strategy in any kind of science.

Hawking's preference for closed universes mirrors the view of Einstein, who always found the idea of a closed universe, a finite construction with no boundary, more appealing to his sensibilities. At the bottom of all this, there seems to be little more than a vague feeling that finiteness is nicer than infiniteness. By contrast, there have always been some physicists and cosmologists who have found the endless stretch of space and time in an infinite universe more comfortable, precisely because of the limitless possibilities it seems to allow. For all of quantum cosmology's sophistication, it ends up being a technically complicated piece of machinery that does little more than offer an incomplete physical justification for what remains an essentially aesthetic preference.

It is idle to place bets on which if any of the current versions of string theory or quantum cosmology will emerge as the precursor of a theory of everything. But is it reasonable even in principle to imagine or hope that the desired theory will tell us everything we want to know? Quantum cosmology is at present more of an idea for a theory than a theory itself. String theory derives from compellingly unique and beautiful mathematical notions, but to get the real world of massive particles and flawed symmetries and separate forces requires the pristine superstring to be adorned with a welter of intricate ornamentation, none of it unique or compelling by any stretch of the imagination. General relativity is supposed somehow to fall out of superstring theory, along with the strong and electroweak interactions, but general relativity by itself then allows an infinite variety of universes and makes no choice between them. What reason is there to suppose that the version of general relativity that comes, if it ever does, from superstrings will have sufficient extra theoretical ornaments attached to it to force the choice of one particular universe, the one we live in, from the infinite range of possibility? Such an outcome is what Einstein hoped for, in dreaming that one day God might be found to have had no choice in the creation of the world; it is what Hawking, evidently, means when he asks, "What place, then, for a creator?"[2]

It may seem like a logical conclusion to the progress of theoretical physics that it should eventually furnish a theory so complete that nothing is left undetermined, but this is an ambition at odds with the spirit of physics since Newton and Galileo. Those who thought that physics was coming to an end a hundred years ago thought so because it seemed that all the rules would soon be known, but they did not ask for more than the rules. The Marquis de Laplace's famous statement that if the position and velocity of all the objects in the world were exactly known at one time then the future and past would be known in infinite detail still left hanging the question of how it all started. Someone had to set the particles in motion; it was not thought that the rules of physics would themselves contain an explanation of what set the ball rolling, only that once the ball was rolling every aspect of its progress would be understandable. Classical physics, even had it been taken to the point of completion imagined by Descartes, Laplace, and many others, was a system of laws only; there was still room for a prime mover. But those who, like Einstein and Hawking, hanker after the more extreme modern version of the end of physics wish to exclude all contingencies from their calculations. The theory of everything should comprise not just the rules of interaction between particles but also boundary conditions for the application of those rules; it is supposed to dictate, of its own accord, the nature of the universe that the theory of everything will inhabit. It is a nice question for philosophers what sort of void this self-starting theory exists in, before it manufactures a universe in which it proceeds to make physics happen.

That the world can by understood by pure reason, not by experimentation but by mentation alone, is a very old idea, reaching back to the ancient Greeks. For the modern advocates of theories of everything, as for the ancient Greeks, reason, logic, and physics are supposed to constitute the unmoved mover, the uncaused effect.

Not every physicist finds this philosophy reasonable or congenial; to some it is the utmost hubris to imagine that a mathematical theory, no matter how exquisite, can contain the answers to every conceivable question. But neither do these skeptics want to face the possibility that physics and the universe as a whole could be the way they are arbitrarily, because some number in the heart of all the complex symmetry breakings of particle physics happens to be five and not six. It may be reckless to suppose that physics can be as complete and encompassing as Einstein and Hawking would like, but it seems positively inglorious that after centuries of effort physics would have to end like a just-so story. The universe is the way it is because it is not any other way; there are four apparently distinct forces of nature because there are neither three nor five.

Between the hopes for a perfect theory of everything and mere arbitrariness lies a compromised philosophy by the name of the anthropic principle. This philosophy begins with the straightforward observation that for us to be able to learn about physics and cosmology we must ourselves have come into existence during cosmological history. It may be that many different universes are allowed by the laws of physics, but perhaps almost all the theoretically possible universes will be inimical to the development of intelligent life. At a basic level, this means that the universe cannot be, for example, only a million years old, because a million years is not enough time for stars, planets, and life to develop. The anthropic principle says that it is no surprise that the universe has properties commensurate with our existence, because if it didn't, we wouldn't exist, and wouldn't be worrying about the matter. Inflation, to this way of thinking, is not needed to explain why the universe should be about ten billion years old; rather, the universe has to have at least that sort of age in order to allow us to live in it and contemplate its age.

This sounds dangerously close to tautology. There is a weak anthropic principle, to which it is hard to object, that says that the universe must be such as to allow life to develop in it. Of course, we do not know precisely what conditions are needed for the development of life, since life on Earth is the only example we know, but at a minimum it seems reasonable to guess that stars (energy sources) and planets (colder condensed bodies on which chemical reactions can take place) are both needed. There must therefore be material in the universe out of which stars and planets can grow, and time enough to do the job. But this is not very specific; it amounts only to an after-the-fact rationalization of things we already know, not a prediction of anything.

Going well beyond this is the strong anthropic principle, that says that not just the general cosmological environment but all of physics itself is tuned to the requirements of our existence. The kind of argument involved in this is, for example, that the details of nuclear physics, which derive in turn from the strong and electroweak interactions, have to be such as to allow stars to shine; in principle, one can imagine a revised version of nuclear physics, obtained by adjusting some of its numerical constants, in which some elements now radioactive are made stable, others now stable become radioactive, and the rates of all nuclear reactions between the various elements change. According to this revised nuclear physics, stars might burn so slowly that they could generate only feeble heat, or so quickly that their life would be over in a few thousand years, before life on any surrounding planets had a chance to develop. If the strong anthropic principle is believed, then everything about nuclear physics, and indeed about all of physics, is just so as to produce the universe we live in.

This releases us from the need for a theory of everything to predict exactly the universe we know. Instead, we are allowed to imagine that all kinds of possible universes can spring up from the quantum ectoplasm, each then setting off on its own course of evolution. In each of these universes, supposedly, the way that the theory of everything breaks down into gravity and the strong and electroweak interactions may be different; in one of these universes, for example, an initial stage of symmetry breaking might not occur, so that the strong and electroweak interactions remain identical and never develop into the separate forces that obtain in our universe. Some of these allowable universes will last only a million years before recollapsing; others will contain no matter; some will have nuclear physics that causes stars to burn up in a thousand years; some will have no nuclear physics at all. But in one happy case, everything is right, and we come along in about ten billion years to start asking questions.

This is surely a generous way of thinking: particle physicists have argued for the existence of all kinds of new particles in addition to the ones we know about, in order to allow supersymmetry or supergravity or superstrings to work, but proponents of the strong anthropic principle are multiplying whole universes in order to provide a home for us, and then, moreover, concealing those universes from us so that we can never see the extravagances enacted for our benefit. This illustrates the basic dilemma in all this puzzling. We live in but a single universe, and we are asked to decide whether we live in the one and only universe that has been created, in which case we have to explain why that universe should have properties congenial to our existence, or in one of an infinite variety of universes, in which case the fact that we live in a congenial one becomes a tautology. The problem is that, by definition, we cannot know about universes other than our own; if we knew about them, they would be part of our universe. Whether other universes exist or not is inherently a question to which an empirical answer is impossible, and the strong anthropic principle is really a way of disposing of the question of the origin of the universe by claiming that the answer is in a place we can never see.

One way out of this dilemma has been suggested. We may live in a region of the universe which is relatively uniform over a large volume, but this is no guarantee that the universe beyond our present range of visibility is uniform. As the universe continues to expand, more of it heaves into sight: it is now about ten billion years old, and we can see out to a maximum distance of ten billion light years; when the universe is a hundred billion years old, we will see out to a hundred billion light years. This allows a variation on the idea of an infinite number of universes: imagine that we live in a universe that is infinite but not uniform throughout, and suppose that the manner in which the universe was born allows not just cosmological geometry but physics itself to vary from place to place. In some parts of this infinite extent, matter might be so scarce that stars and planets never form; in some parts, stars might form but, because nuclear physics is different, burn out in a hundred years. The adherent of the anthropic principle then merely observes that we must live in a part of the universe that is amenable to life. There is no proliferation of universes, and indeed if we wait long enough for the universe to expand we should see bits of nonuniform, irregular, and chaotic cosmology coming over the horizon. We live on an island of calm set down purely at random in a turbulent sea.

But this doesn't really work, either. If the only anthropic requirement for such a universe is that it has to provide a place for us to live, there is no reason why so large a region of the universe had to be made smooth, nor why that region had to be made so extremely smooth. It has become an urgent problem for cosmologists to reconcile the existence of large-scale structures— galaxy clusters and so on—with the almost perfect uniformity of the microwave background. We certainly could have evolved and lived comfortably, one would think, in a universe that had less dramatic structures in it, and we would not be anguished greatly by a microwave background somewhat more distorted than the one we observe. Unless the strong anthropic principle is extended to the point of arguing that the universe had not only to give us a physically acceptable place to live but also to provide us lasting intellectual stimulation, the present specifications of the universe, or at least of the bit we live in, seem to be far in excess of what is required.

One faintly plausible way out of the conundrum was offered by the Russian cosmologist Andrei Linde, in his 1983 theory of chaotic inflation.[3] Inflationary theories in general are designed to produce a large smooth universe no matter how irregular the starting conditions, but in any particular inflationary theory it is always possible to find some cosmic initial conditions, most likely very odd ones, that do not lead to inflation at all, and hence do not produce a smooth universe. Even with inflation it remains necessary, therefore, to put some limitations on the state of the pre-inflationary universe, but this is a big detraction from the appeal of inflation; it brings back the old problem of trying to figure out why some initial conditions, the ones that lead to inflation, are chosen over the ones that do not.

But if the universe before inflation is irregular—or chaotic, as Linde termed it—then some regions in it will have conditions that lead to inflation, while others do not. Those regions that inflate will grow at the expense of the ones that do not; the inflated regions will be huge and uniform, and separated from each other by margins into which all the still chaotic bits of the universe that didn't inflate have been squashed. The only habitable regions of the universe are the ones that underwent inflation, and they are so large that the extent of the uniformity of the bit of the universe we live in is to be expected. Whether the anthropic principle is needed or not at this point is almost a matter of taste, since chaotic inflation has demarcated the universe very clearly into regions that are very large and uniform, hence habitable, and small remainders of the initial irregular state, which are not.

In fact this picture is excessively simplified. Even if some parts of the universe inflate a great deal and some not at all, there are bound to be regions that inflate a bit, creating a moderately smooth region of middling extent. Such regions may well be much less numerous than the ones that inflated hugely, but they will exist, and something in addition to the anthropic principle seems be needed to explain why we live in an extremely smooth region, not a moderately smooth one. And if we add to the chaotic model the idea that the theory of everything may break down into different particle-physics theories (different nuclear physics, different relative strengths of weak and electromagnetic interactions, and so on) from place to place, then the chaotic inflationary model falls apart. Some regions, as before, will inflate enormously, and some initially irregular regions will not inflate. But in addition there will be large regions, some fairly uniform, others not uniform at all, that don't inflate because particle physics in those regions did not develop from the initial theory of everything in such a way as to allow inflation. Now we are back to the same problem: if, out of an initially irregular cosmos, regions develop of varying uniformity and varying size, why should we live in a region that is so very uniform over so very large an extent? The anthropic principle is simply insufficient to explain this.

The anthropic principle begins to look like a name for the repository in which we set aside all the things that physics cannot yet explain. It is always a last resort. Physics is used, extrapolated as far as it can be in order to explain why the universe has to be the way it is, but when this is done and some facts remain unexplained, the anthropic principle is brought in to select, from the limited menu that physics has offered, the universe that seems most agreeable to us. Necessarily, the scope of the anthropic principle depends on how far physics has advanced. The anthropic principle is used to explain those things for which physics alone, we suspect, cannot provide an answer. It performs the role that less artful scientists in earlier ages ascribed unabashedly to a prime mover, or to God.

If we discard the anthropic principle, we are back to the original dilemma. Either we believe that a true theory of everything will one day come about, with the capacity to answer every question we can put to it, or we must accept an irrreducible element of arbitrariness in the making of the universe and the specification of particle physics. The latter course seems a painfully inadequate end to the great quest that is theoretical physics, so perhaps we must believe that the hope for a true theory of everything, in the most extreme sense possible, is justified.

Can this hope be sustained? Present attempts at theories of everything rely on an abundance of fundamental principles (which themselves may or may not be independently testable) and suffer at the same time from a deficiency of detail: the theories must be augmented with compactification of extra dimensions, symmetry breakings to give masses to some particles and not others, more symmetry breakings to distinguish the various particle interactions from each other, and so on. This ornamentation does not emerge naturally in any of the theories of everything that have been talked about so far, and all of it has to be added in by hand, to make the theory come out the way we need it to come out. But at the same time, these details, these departures from the perfection of the theory of everything in its pristine state, are the only things we can hope to measure. Proton decay, for example, is little more than a side effect, a tiny aftershock of grand unification, but we cannot reach into the heart of grand unification by direct experiment, and so we must search for proton decay as a way of testing if grand unification is a good theory. But even if proton decay is detected, we will have verified only one small ornament of the theory, not the whole thing.

It is a characteristic of all would-be theories of everything that whatever mathematical and aesthetic perfection they possess is apparent only at the ultrahigh temperatures of the first moments of cosmic history. Once the big bang is under way, and the symmetries of the theory of everything begin to fracture into what we now observe, direct sight of that theoretical perfection is lost. We can study only the universe around us—the shattered, complicated, messy remnant that is our sole bequest from the theory of everything—like archaeologists trying to reconstruct the daily habits of the Babylonians from a handful of shards.

All the easy experiments in particle physics have been done. Scientists can no longer build more powerful accelerators and find new particles by the bucketful. It took a decade of planning and political fighting to build the machine at CERN that eventually captured the W and Z particles and thus went part way toward verifying the unified electroweak theory. The Superconducting Supercollider was a decade in planning, and only two years ago was the first concrete poured. It will be another decade before the machine is up and running; then, perhaps, it will find the Higgs boson of electroweak unification and thereby, at a cost of ten billion dollars and twenty years' hard labor, provide a little more evidence in favor of electroweak unification. The particle detectors for machines like the supercollider are billion-dollar projects, and consume the efforts of many graduate students and young researchers, whose reward is to be one of two or three hundred authors on a four-page paper published in the *Physical Review Letters*.

The search for proton decay has so far taken similar outlays of time, money, and effort. In abandoned mines, tanks containing thousands of tons of ultrapure water have been installed and surrounded by detectors, waiting to trap the telltale particles from the decay of one proton in one of the water molecules in the thousands of tons of water. A decade has gone by, and no proton decays have been seen; the only way to go on with such an experiment is to build a bigger tank or simply wait longer. One begins to imagine physics experiments being passed on from one generation of researchers to another, waited upon and watched with stubborn patience in case, one day, something unusual should come along. Particle physicists have sometimes drawn an analogy between their machines and the Gothic cathedrals of medieval Europe—both, in their own ways, monuments to a search for truth by their respective communities. The analogy may be more apt than physicists would like: Gothic cathedrals occupied generations of now nameless carpenters and masons.

Still, the Gothic cathedrals were built, and remain today. Physicists may feel that if the world is prepared to foot the bill, they are prepared to embark on efforts that they will not see finished in their lifetimes. As long the universe exists, and keeps on expanding, they can keep doing experiments. If the universe is closed, and eventually ends in a "big crunch," physics experiments will end too, but if the universe is open it has an infinite future, and it might seem that physicists must have an infinite time to complete their progress toward a theory of everything.

The future may be infinite indeed, but it may also be empty. Our Sun will not last forever, but when it dies and consumes our Earth in the process, we will have perhaps found a younger sun elsewhere, with hospitable planets that we can live on. New stars are forming even now from the remnants of old ones, their material dispersed into space by supernovae. As long as we can hop from star to star, we can live in this galaxy for quite some time yet.

But thermodynamics, as Clausius realized a hundred years ago, is relentless and unforgiving. Each generation of stars uses up forever a portion of hydrogen, the nuclear fuel with which the universe was endowed; there is not a limitless supply. But there may be other ways to mine energy from the cosmos and thereby sustain our existence. Dead stars in our own galaxy may turn into black holes, which draw into them gas, dust, and smaller stars floating around in interstellar space. As this material is sucked down the black hole, it forms an inward-spiraling disk of hot gas—hot because its inexorable fall into the black hole releases gravitational energy, in the way Lord Kelvin imagined that the Sun gave off heat by pulling down meteorites. One day there will be no new stars in the galaxy, because there will be no nuclear fuel left for them to run on, but by that time we can stay alive on dead planetary systems that are drifting toward a giant black hole at the center of the galaxy, hopscotching outward as each temporary habitation is sucked down.

In the heat-death imagined by Clausius at the end of the last century, the universe was static, and as all its sources of energy were used up, it filled uniformly with heat until everything was at the same lukewarm temperature; physical processes were suffocated. But our universe is expanding, and though the constant extraction of energy from one source or another generates heat, that heat is constantly diluted by the expansion of space, just as the photons in the microwave background have been reduced to a mere three degrees above absolute zero and continue to cool. Heat is constantly generated, but the universe cools nevertheless, and in our distant future we face death by freezing.

Eventually, the universe will be nothing but a collection of dead black holes in a tenuous sea of discarded radiation whose energy is being steadily diminished by the expansion. All is not quite over yet: black holes evaporate, as Hawking realized, slowly losing mass by giving off energy and particles. Black holes of galactic mass, trillions of times the mass of the Sun, will radiate photons of hopelessly low energy and will take eons to evaporate altogether, but in their death throes, when they are very tiny, they will give up all their remaining mass in a burst of high-energy radiation and particles. Here is another usable source of energy, perhaps, if we can stick around long enough to get the benefit of it. But if grand unification is correct, matter itself, based on protons and neutrons, will not survive forever; the only absolutely stable particles, apparently, are electrons and positrons, along with photons, neutrinos, and gravitons. This is weak material from which to make a habitable universe. In the end, therefore, nothing will survive except feeble photons, undetectable neutrinos, and perhaps a few electrons and positrons so far apart that they will never meet and annihilate each other. This is the revised "heat death" of the universe, which imposes not stifling by accumulated heat but a slow diminution of activity and a decay of matter, one proton at a time.

Even if we live in an infinite universe, therefore, there is not infinite time available to us, either for our experiments or our imaginations. As energy becomes harder to acquire, it is unlikely that anyone will want to squander it on huge experimental devices aimed at determining the value of the next decimal place of some parameter in a theory of fundamental physics. Even thinking about theoretical physics uses up energy, and contributes to the death of the universe; every thought process in the brain is a thermodynamic action, and contributes to the growing entropy of the universe.

We are already at the point where experiments are becoming impossible for technological reasons and unthinkable for social and political reasons. An accelerator bigger than the supercollider would be a vast technical challenge, and even if physicists are willing to try it, the likelihood of society paying the bills seems faint. Thermodynamics, however, is stronger than any political force; as the universe winds down, energy will become unavailable at any cost, and physics itself, in the form of the venerable strictures of thermodynamics, will make it impossible for physicists to do any but a tiny fraction of the experimental work that would be needed to test a theory of everything.

The physicists must hope instead that they can complete physics in the manner the ancient Greeks imagined, by means of thought alone, by rational analysis unaided by empirical testing. The ultimate goal in physics seems to demand, paradoxically, a return to old ways. Modern physics was set on its present course by the pragmatic methods of Newton and Galileo and their many successors, and this effort, three centuries old, has led to the elaborate physical understanding we now possess. But this kind of physics seems to have run its course. Experiments to test fundamental physics are at the point of impossibility, and what is deemed progress now is something very different from what Newton imagined. The ideal of a theory of everything, in the minds of the physicists searching for it, is a mathematical system of uncommon tidiness and rigor, which may, if all works out correctly, have the ability to accommodate the physical facts we know to be true in our world. The mathematical neatness comes first, the practical explanatory power second. Perhaps physicists will one day find a theory of such compelling beauty that its truth cannot be denied; truth will be beauty and beauty will be truth—because, in the absence of any means to make practical tests, what is beautiful is declared ipso facto to be the truth.

This theory of everything will be, in precise terms, a myth. A myth is a story that makes sense within its own terms, offers explanations for everything we can see around us, but can be neither tested nor disproved. A myth is an explanation that everyone agrees on because it is convenient to agree on it, not because its truth can be demonstrated. This theory of everything, this myth, will indeed spell the end of physics. It will be the end not because physics has at last been able to explain everything in the universe, but because physics has reached the end of all the things it has the power to explain.

NOTES

Prologue: The Lure of Numbers

The work and thought of Albert Einstein cast a long shadow over twentieth-century physics. There are numerous biographies of him. Ronald W. Clark's *Einstein: The Life and Times* (New York: Avon, 1971) is a voluminous narrative, but skimps the science. Abraham Pais's *Subtle Is the Lord* ... (Oxford: Oxford University Press, 1982) presents a more thorough but more technical analysis of Einstein's development as a scientist.

The scientific ideas touched on in this chapter are for the most part described more fully in the rest of the book, and references are given in notes for later chapters.

1. Reprinted in *The World Treasury of Physics, Astronomy, and Mathematics*, ed. Timothy Ferris (Boston: Little, Brown, 1991), pp. 526–540.
2. Albert Einstein, *Sidelights on Relativity* (New York: Dover, 1983), p. 28.
3. Albert Einstein, *Ideas and Opinions* (New York: Bonanza Books, 1954), p. 292.
4. Max Planck, *Where Is Science Going?* (Woodbridge, Conn.: Ox Bow Press, 1981), p. 214.
5. Richard P. Feynman, *QED: The Strange Theory of Light and Matter* (Princeton, N. J.: Princeton University Press, 1985), p. 129.
6. Paul Ginsparg and Sheldon L. Glashow, "Desperately Seeking Superstrings," *Physics Today*, May, 1986, pp. 7–9. The authors express concern that theoretical physics is moving into a world of unverifiable mathematical invention. Similar concerns are

expressed by Glashow and R.P. Feynman, in interviews published in *Superstrings: A Theory of Everything?* (Cambridge: Cambridge University Press, 1988), ed. Paul C. W. Davies and John Brown.

Chapter 1: Lord Kelvin's Declaration

The saga of Kelvin's battle with the geologists over the age of the Earth, and of its salutary effect on geological theorizing, is recounted by Joe B. Burchfield in *Lord Kelvin and the Age of the Earth* (New York: Science History Publications, 1975). A full but technical account of the development of electromagnetic theory from the early days into the first part of the twentieth century is by Edmund T. Whittaker in *A History of the Theories of the Aether and Electricity* (New York: Dover, 1989; two volumes reprinted as one).

For aspects of the development of thermodynamics and statistical mechanics, and of the people who contributed to these subjects, some useful references are: Engelbert Broda, *Ludwig Boltzmann: Man, Physicist, Philosopher*, trans. Larry Gay and Engelbert Broda (Woodbridge, Conn.: Ox Bow Press, 1983); Ivan Tolstoy, *James Clerk Maxwell* (Edinburgh: Canongate, 1981); and John G. Crowther, *British Scientists of the Nineteenth Century* (London: Routledge, 1962).

1. Henry Adams, *The Education of Henry Adams* (New York: American Heritage Library, Houghton Mifflin, 1961), p. 401. Adams mentions an age limit of twenty million years (an earlier and more stringent calculation by Kelvin) rather than the later estimate of one hundred million years.
2. Burchfield, *Lord Kelvin and the Age of the Earth*, p. ix.
3. Pierre-Simon de Laplace, *The Analytical Theory of Probability*, in *The World of Physics*, ed. Jefferson H. Weaver (New York: Simon and Schuster, 1987), vol. 1, p. 582.
4. Whittaker, *A History of the Theories of Aether and Electricity*, vol. 1, p. 255.
5. Burchfield, *Lord Kelvin and the Age of the Earth*, p. 164.

Chapter 2: Beyond Common Sense

A careful and profound account of special relativity is by Hermann Bondi, *Relativity and Common Sense* (London: Heinemann, 1965). The early development of the quantum theory is simply related by George Gamow in *Thirty Years That Shook Physics: The Story of Quantum Theory*

(New York: Dover, 1985) and by Emilio Segrè in *From X-rays to Quarks* (San Francisco: W.H. Freeman, 1980). A more sophisticated and technical account is in the early chapters of *Inward Bound*, by Abraham Pais (Oxford: Oxford University Press, 1965).

Max Planck, in *Where Is Science Going?* (Woodbridge, Conn.: Ox Bow Press, 1981), describes some of his views on the meaning of quantum mechanics, as does Werner Heisenberg in various collections of essays: *Physics and Philosophy* (London: Penguin, 1989), *Physics and Beyond* (New York: Harper and Row, 1971), and *Across the Frontiers* (Woodbridge, Conn.: Ox Bow Press, 1990).

1. From the scholium to *Philosphiae naturalis principia mathematica*, quoted, for example, in *Theories of the Universe*, ed. Milton K. Munitz (Glencoe, Ill.: The Free Press, 1957), p. 202.
2. Pais, *Inward Bound*, p. 189.
3. Gamow, *Thirty Years That Shook Physics*, p. 81.
4. Heisenberg, *Physics and Philosophy*, p. 29.

Chapter 3: The Last Contradiction

The use of the Hulse-Taylor binary pulsar in tests of general relativity has been described by Clifford M. Will in *Was Einstein Right?: Putting General Relativity to the Test* (New York: Basic Books, 1986). Will also discusses other ways in which Einstein's theory has been addressed by experimental physicists.

Quantum Profiles (Princeton, N. J.: Princeton University Press, 1991), by Jeremy Bernstein, offers an affectionate description of John Bell, who died in 1990. The most important of Bell's work on the meaning of quantum theory has been published in a single volume, *Speakable and Unspeakable in Quantum Mechanics* (Cambridge: Cambridge University Press, 1987).

1. David Bohm, *Quantum Theory* (Englewood Cliffs, N. J.: Prentice-Hall, 1951).
2. John S. Bell, "On the Einstein–Podolsky–Rosen Paradox," reprinted in *Speakable and Unspeakable in Quantum Mechanics*, pp. 14–21.
3. John C. Polkinghorne, *The Quantum World* (Princeton, N. J.: Princeton University Press, 1984). See particularly the discussion in chapter 8.
4. Quoted by Jeremy Bernstein in *Quantum Profiles*, p. 137.

Chapter 4: Botany

A detailed and accessible narrative account of the development of particle physics, from the beginnings of quantum mechanics to electroweak
unification, is *The Second Creation*, by Robert P. Crease and Charles C.
Mann (New York: Macmillan, 1986). A shrewder but much more technical history, with a stronger sense of the growth of ideas, is *Inward Bound*,
by Abraham Pais (Oxford: Oxford University Press, 1985). Also useful
are *The Tenth Dimension*, by Jeremy Bernstein (New York: McGraw Hill,
1989), and the later chapters of *From X-Rays to Quarks*, by Emilio Segrè
(San Francisco: W. H. Freeman, 1980).

1.   Werner Heisenberg, *Physics and Philosophy* (London: Penguin, 1989),
      p. 60.
2.   Crease and Mann, *The Second Creation*, p. 169.
3.   Ibid., pp. 305–7.

Chapter 5: The Expanding Universe

A contemporary account of early days in cosmology, including the identification of distant galaxies and the discovery of their redshifts, is Edwin
P. Hubble's *The Realm of the Nebulae* (1935; reprint, New Haven: Yale University Press, 1985). Also of historical interest is a descriptive account of
astronomy in the 1920s by James Jeans, *The Universe Around Us* (London:
Macmillan, 1929). A selection of cultural and scientific essays was collected by Milton K. Munitz in the volume *Theories of the Universe* (Glencoe, Ill.: The Free Press, 1957). Reminiscences by some of the founders of
the subject are in *Modern Cosmology in Retrospect*, edited by Bruno Bertotti,
Roberto Balbinot, Silvio Bergia, and Andrea Messina (Cambridge: Cambridge University Press, 1990); the volume also includes essays by contemporary analysts on historical matters.

A good up-to-date pedagogical account of the present state of cosmology is Joseph Silk, *The Big Bang* (San Francisco: W. H. Freeman, 1980).
Ancient Light, by Alan Lightman (Cambridge: Harvard University Press,
1991), and *Coming of Age in the Milky Way*, by Timothy Ferris (New York:
Morrow, 1988), both recount the history of modern cosmology.

1.   Hubble, *The Realm of the Nebulae*, p. 122.
2.   Ralph A. Alpher and Robert Herman, "Big Bang Cosmology and
      Cosmic Blackbody Radiation," in *Modern Cosmology in Retrospect*, pp.

129–157. This is an account of the origin of the modern idea of the big bang, by two of its three inventors.

3. Robert W. Wilson, in "Discovery of the Cosmic Microwave Background" in *Modern Cosmology in Retrospect*, pp. 291–307, gives the historical perspective of one of the discoverers.

4. See the account by Michael Riordan and David N. Schramm in *The Shadows of Creation: Dark Matter and the Structure of the Universe* (San Francisco: W. H. Freeman, 1990), pp. 86–90.

Chapter 6: Grand Ideas

Dennis Overbye, in *Lonely Hearts of the Cosmos* (New York: HarperCollins, 1991), describes cosmologists and their achievements; the sometimes awkward relationship between particle physicists and cosmologists is a theme of the latter part of his book. Timothy Ferris, in *Coming of Age in the Milky Way* (New York: Morrow, 1988), and Alan Lightman, in *Ancient Light* (Cambridge: Harvard University Press, 1991), also talk about the recent impact of particle physics on cosmology.

1. Edward P. Tryon, "Is the Universe a Vacuum Fluctuation?" *Nature*, December 14, 1973, p. 396.

2. Alan H. Guth, "Inflationary Universe: A Possible Solution to the Horizon and Flatness Problems," *Physical Review D*, 1981, vol. 23, p. 347.

3. Sheldon L. Glashow and Ben Bova, *Interactions* (New York: Warner, 1988), p. 207.

Chapter 7: Dark Secrets

Dennis Overbye's *Lonely Hearts of the Cosmos* (New York: HarperCollins, 1991) describes many of the developments mentioned here. *The Shadows of Creation: Dark Matter and the Structure of the Universe*, by Michael Riordan and David N. Schramm (San Francisco: W. H. Freeman, 1990), provides a full account of modern theories of galaxy formation.

Supersymmetry is described at a semi-technical level in the essay "Is Nature Supersymmetric?" by Howard E. Haber and Gordon L. Kane, in *Scientific American*, June, 1986.

1. Allan Sandage is quoted in *Lonely Hearts of the Cosmos*, p.40, as saying that "Bondi's first theorem is that whenever there is a conflict

between a well-established theory and observations, it's always the observations that are wrong."

2. *Nature,* February 21, 1985, p. 626.
3. *Theories of the Universe,* ed. Milton K. Munitz (Glencoe, Ill.: The Free Press, 1957), p. 211.
4. The "Death of Cold Dark Matter" was announced on the front cover of *Nature,* January 3, 1991, in connection with the article by Will Saunders et al. pp. 32–38. The front cover was an editorial idea and did not entirely please the authors of the article.

Chapter 8: March of the Superlatives

Supersymmetry, supergravity, and superstrings are all so recent that few simple guides to them exist. *Superforce,* by Paul C. W. Davies (New York: Simon and Schuster, 1984), and *Fearful Symmetry,* by Anthony Zee (New York: Macmillan, 1986), both describe supersymmetry and supergravity.

For superstrings there is the very clear article "Superstrings" by Michael B. Green in *Scientific American,* September, 1986, as well as more technical reviews such as "The Superstring: Theory of Everything or of Nothing?" by John Ellis, *Nature,* October 16, 1986, pp. 595–597, and "Unification of Forces and Particles in Superstring Theories," by Michael B. Green, *Nature,* April 4, 1984, pp. 409–414.

1. Hawking's lecture is reprinted as an appendix to John Boslough, *Stephen Hawking's Universe* (New York: Avon, 1985).
2. Daniel Z. Freedman and Peter van Nieuwenhuizen, *Scientific American,* February, 1978, p. 126.
3. See, for example, the article "The Quest for a Theory of Everything Hits Some Snags," by Faye Flam, *Science,* June 12, 1992, pp. 1518–1519.

Chapter 9: The New Heat Death

The classical heat death of the universe, and Ludwig Boltzmann's thoughts on it, are in Engelbert Broda's *Ludwig Boltzmann: Man, Physicist, Philosopher* (Woodbridge, Conn.: Ox Bow Press, 1983). Karl Popper also has some comments on the subject in chapter 35 of his autobiographical *Unended Quest* (London: Fontana, 1976).

Stephen Hawking's work on quantum cosmology is described in his inimitable way in *A Brief History of Time* (New York: Bantam, 1988). An exhaustive account of the history and significance of the anthropic principle is John D. Barrow and Frank J. Tipler, *The Anthropic Cosmological Principle* (Oxford: Oxford University Press, 1986).

1. See especially chapter 8 ("The Origin and Fate of the Universe") of *A Brief History of Time*.
2. Ibid., p. 141.
3. For a description of chaotic inflation by its author, see "Particle Physics and Inflationary Cosmology," by Andrei D. Linde, *Physics Today*, September, 1987, pp. 61–69.

INDEX

Space (*continued*):
    density of, 157–58; extra dimen-
    sions of, 19, 189–90, 219–22,
    224–26, 227–30, 250; as flat, 137,
    139, 141–42, 143, 147; and the
    "heat death" of the universe,
    236–37; and high-frequency radi-
    ation, 63–64; and steady-state
    theory, 144; and string theory,
    224; two-dimensional vs. three-
    dimensional, 140–42
Space shuttles, 83–84, 85
Spite, François, 152
Spite, Monique, 152
Stanford Linear Accelerator Center,
    116, 154
Statistical mechanics, 33–36, 41, 91,
    93, 232; and the radiation spec-
    trum, 64–65
Steady-state theory, 143–46, 151;
    and big bang theory, 152; and the
    Hubble constant, 183; and the
    matter-antimatter symmetry,
    159, 160
Steam engines, 30, 31, 33–35, 37,
    46, 232
Steinhardt, Paul, 179
Strangeness, concept of, 114
String theory, 19, 222–31, 247; and
    the definition of strings, 223; and
    the "heat death" of the universe,
    235–37, 240; and quantum cos-
    mology, 244; and superstrings,
    222–31, 235, 240
Subjectivity, 54
Sun, 28, 89; age of, 29; and black
    holes, 239, 254; eclipse of, 11, 12,
    86–87; and the Michelson-Morley
    experiment, 43–44, 55; orbit of
    Earth around, 43–44, 55, 133–34.
    See also Solar System
Superconducting Supercollider, 23,
    113, 124, 155, 173, 252
Supernovae, 108, 253
Superparticles, 195
Symmetry, 1, 5, 120, 123, 170–72,
    250; and dark matter, 187–88;
    and the division of the elemen-

tary world into fermions and
    bosons, 189, 190; "global," 214;
    and grand unification, 163, 174,
    178; matter-antimatter 159–61,
    163–64, 166–68, 181–82, 229;
    "mirror," 226–27; and the notion
    of an expanding universe, 181;
    and quantum cosmology, 244;
    and string theory, 226; super-,
    188, 190–96, 213–15, 219–22,
    226, 228–30, 247
Synchrocyclotron, 112–13

Tau neutrinos, 121, 153, 219
Tauons, 121, 122, 153, 188
Tautology, 246, 248
Taylor, Joseph, 88–90
Taylor, Richard, 116
Temperature, 45–47, 53, 63–64, 250;
    and black holes, 238, 239; of an
    expanding universe, 129, 148–50,
    151–52, 153, 180; and galaxy for-
    mation, 199; of the microwave
    background, 158, 202–4, 205;
    and solar energy, 233; and vacu-
    ums, 175. *See also* Thermodynam-
    ics
Tevatron accelerator (Fermi National
    Accelerator Laboratory), 23,
    101–2, 105, 112–13, 118, 155
Theology, 51–52, 53, 235. *See also*
    God
Thermodynamics, 41; and atomism,
    34; and color and temperature,
    46–47; and the "heat death" of
    the universe, 232–55; and Kelvin,
    28, 33; laws of, description of,
    30–31; and the microwave back-
    ground, 155; and the notion of an
    expanding universe, 147–48;
    original conception of, 30–33;
    and the radiation spectrum,
    64–65; and statistical mechanics,
    33–35, 36; universality of, 33
Thomson, William. *See* Kelvin
    (William Thomson)
Time, 58–62; as absolute, 58–59, 60,
    61, 75, 77, 79, 80–81; and the